Shane
Brubaker

THE SYNAPTIC
ORGANIZATION
OF THE BRAIN

THE SYNAPTIC ORGANIZATION OF THE BRAIN

AN INTRODUCTION

Gordon M. Shepherd M.D., D.Phil.

ASSOCIATE PROFESSOR OF PHYSIOLOGY
YALE UNIVERSITY SCHOOL OF MEDICINE

New York
Oxford University Press
London 1974 Toronto

PREFACE

This book is based on a seminar course that I gave while a visiting professor of the Institute of Neurological Sciences at the University of Pennsylvania in 1971-1972. I thank Dr. Eliot Stellar for the hospitality of the Institute, and Dr. James Sprague and especially Dr. Sol Erulkar for unstinting support throughout the course.

The participants in the course represented a wide range: undergraduate and graduate students, postdoctoral workers, and faculty. I thought this was as it should be. The students forced me to simplify and make comparisons; my colleagues forced me to give important details their due, and make principles relevant to research interests. In developing and extending the course material into this book, I have kept this wide audience constantly in mind.

When it has come down to making the necessary compromises, I have been acutely mindful of my fellow research workers. No one can be more aware than I am that this book, and each section within it, is only an introduction to the matters at hand. For oversimplifications and distortions of complex problems, and, alas, omissions of the work of esteemed colleagues, I beg forgiveness.

I regret other shortcomings. Neuropharmacology deserves more emphasis. Studies of the developing nervous system are here lacking; the present account gives too static a representation of synaptic organization as it is encountered in the adult. And far more could be said about relations to behavior. But the book is long enough as it stands. I hope that it will help to ·define synaptic organization as a subject that is, to some extent, sufficient unto itself, and that will serve as a foundation for the study of the other aspects of brain function and of other parts of the nervous system.

In recording my debts, I am happy to begin with my mentors, Professor Charles G. Phillips, F.R.S., Dr. Wilfrid Rall, and Professor David Ottoson. I have been extraordinarily fortunate in having them as teachers, colleagues, and friends. I have also been very fortunate in my collaborators, Drs. T. P. S. Powell, T. S. Reese, M. W. Brightman, and J. S. McReynolds, and in my own students, Drs. L. B. Haberly, L. J. Land, T. V. Getchell, and J. S. Kauer. Many other colleagues have given me valuable help and advice along the way, and many have been kind enough to read chapters of the book and offer detailed criticisms. I used to think it something of a cliché that an author gives credit to his colleagues for the merits of his book; now I realize the truth of that statement. Certainly the shortcomings are entirely mine.

My work has been supported by a succession of postdoctoral fellowships, special fellowships, and research grants from the United States Public Health Service and the National Institutes of Health. I am grateful for this support in an area—the olfactory—that for many years seemed to have little relevance beyond its own confines. If this serves as another example of the importance of basic research in little-known areas, the book will have been additionally useful.

My departmental chairmen—Professors C. C. Hunt, G. H. Giebisch, J. F. Hoffman—have given me generous support, including the relief from administrative chores that made writing possible. Mr. Jeffrey House, of Oxford University Press, has been instrumental in initiating this project and seeing it through. My wife, Grethe, has contributed in many ways, not least by her resolve that this book could be written.

G. M. S.

Yale University
New Haven, Connecticut
May, 1974

CONTENTS

1. INTRODUCTION   3

   Traditional Concepts of Synaptic Organization, 4

   Synaptic Organization of Local Regions, 7

   Plan of the Book, 9
      Dimensions, 11.

Part 1
PRINCIPLES OF STRUCTURE AND FUNCTION

2. AXONS, DENDRITES, AND SYNAPSES   15

   Parts of the Neuron, 15
      *Cell Body, 18; Axon Hillock, 20; Initial Segment, 20;
      Axon, 21; Axonal Branches, 21; Dendrites, 22; Termi-
      nals, 24.*

   Types of Synapses, 26
      *Membrane Juxtapositions, 26; Membrane Appositions,
      26; Chemical Synapses, 28; Patterns of Synaptic Con-
      nections, 31; Vesicles, 33.*

   Summary, 34

3. ACTION POTENTIALS AND SYNAPTIC
   POTENTIALS   35

   Membrane Potential, 35

   Action Potential, 37
      *Action Potentials in Thin Axons, 41; Action Potentials
      in Dendrites, 43.*

Synaptic Potentials, 44
Electrical Synapses, 45; Chemical Synapses, 46; Synaptic Integration, 48; Transmitter Mechanisms, 50; Dale's Principle, 53; Transmitter Mechanisms in Central Synapses, 54.

4. DENDRITIC ELECTROTONUS   58

Steady-State Electrotonus, 59
Dendritic Length, 62; Dendritic Diameter, 63; Dendritic Branching, 66.

Transient Electrotonus, 69
Synaptic Interactions, 73.

Part 2
ORGANIZATION OF NEURONAL SYSTEMS

5. SPINAL CORD: VENTRAL HORN   79

Neuronal Elements, 81
Inputs, 81; Principal Neuron, 83; Intrinsic Neurons, 85; Neuronal Populations, 87.

Synaptic Connections, 89

Basic Circuit, 92

Synaptic Actions, 96
Recurrent Inhibition, 100.

Dendritic Properties, 102
Synaptic Integration, 106; Types of Inhibition, 109.

6. OLFACTORY BULB   111

Neuronal Elements, 113
Inputs, 113; Principal Neuron, 114; Intrinsic Neurons, 116; Cell Populations, 118.

Synaptic Connections, 118
Glomerular Layer, 119; External Plexiform Layer, 120; Granule Layer, 121; Glia, 122.

Basic Circuit, 122

Synaptic Actions, 128
Dendrodendritic Recurrent Inhibition, 131; Rhythmic Activity, 133.

Dendritic Properties, 134
Mitral Cell, 134; Granule Cell, 139; Periglomerular Cell, 143.

7. RETINA   145

Neuronal Elements, 146
Inputs, 147; Principal Neuron, 148; Intrinsic Neurons, 149; Cell Populations, 153.

Synaptic Connections, 155
Outer Plexiform Layer, 155; Inner Plexiform Layer, 158.

Basic Circuit, 159

Synaptic Actions, 164

Dendritic Properties, 169
Retinal Receptors, 170; Bipolar Cells, 172; Horizontal Cells, 174; Amacrine Cells, 175; Ganglion Cells, 176.

8. CEREBELLUM   179

Neuronal Elements, 180
Inputs, 180; Principal Neuron, 183; Intrinsic Neurons, 184; Cell Populations, 187.

Synaptic Connections, 188

Basic Circuit, 193
Deep Cerebellar Nuclei, 199.

Synaptic Actions, 199
Deep Cerebellar Nuclei, 205.

Dendritic Properties, 207
Active Properties, 207; Dendritic Inhibition, 210; Dendritic Spines, 211.

9. THALAMUS   214

Neuronal Elements, 216
Inputs, 217; Principal Neuron, 219; Intrinsic Neurons,

220; Cell Populations, 221.

Synaptic Connections, 222

Basic Circuit, 225

Synaptic Actions, 229
    Inhibitory Pathways, 230; Rhythmic Activity, 232.

Dendritic Properties, 233
    Principal Neuron, 233; Intrinsic Neurons, 236.

10. OLFACTORY CORTEX   238

Neuronal Elements, 240
    Input, 240; Principal Neuron, 241; Intrinsic Neurons, 243; Cell Populations, 244.

Synaptic Connections, 245

Basic Circuit, 247

Synaptic Actions, 250
    Recurrent Inhibition, 252; Rhythmic Activity, 252.

Dendritic Properties, 254

11. HIPPOCAMPUS   259

Neuronal Elements, 261
    Inputs, 262; Principal Neuron, 263; Intrinsic Neurons, 265; Cell Populations, 266; Dentate Fascia, 267.

Synaptic Connections, 268

Basic Circuit, 272

Synaptic Actions, 277
    Inhibitory Pathways, 280; Rhythmic Activity, 281; Dentate Fascia, 283.

Dendritic Properties, 284
    Electrical Parameters, 284; Fast Prepotentials, 285.

12. NEOCORTEX   288

Neuronal Elements, 290
    Input, 290; Principal Neuron, 291; Intrinsic Neurons, 295.

Cortical Cytoarchitectonics, 297

Cell Populations, 299

Synaptic Connections, 300

Basic Circuit, 303

Synaptic Actions, 310

Afferent Excitatory and Inhibitory Actions, 310; Recurrent Excitatory and Inhibitory Actions, 313; Sensory Stimuli, 314; Motor Cortex, 317; Rhythmic Potentials, 319.

Dendritic Properties, 321

Electrical Parameters, 321; Synaptic Transfer and Lamination, 322; Dendritic Spines, 325.

REFERENCES   329

INDEX   353

·

THE SYNAPTIC
ORGANIZATION
OF THE BRAIN

1

INTRODUCTION

Most organs of the body are made up of cells that are relatively simple in form and either similar or, at least, obviously complementary in function. That the brain is organized in this way is by no means evident; the impression is rather the reverse. The brain seems to be made up of a bewildering complexity of parts, and the cells within the parts seem to be characterized by an inscrutable complexity of form, extent, and relationships with each other. This is true of the nerve cells, or *neurons*, which transmit nervous signals, as well as of the surrounding cells, or neuroglia, and the many types of sensory, muscle, and gland cells to which the neurons are functionally related.

The student of the brain is therefore faced with a severe challenge: What are the principles that govern the organization of neurons in the different parts? If the subject is to be a science, it must be built on this foundation. If neurons are indeed organized according to some common principles, are our present methods sufficient to identify them?; do these principles provide meaningful insights into the nature of brain functions?; and do they provide a useful basis for experiment and theory?

The function of any organ depends upon interactions between its constituent cells, and in the brain, interactions between neurons are the very essence of its function. These interactions take place through connections termed *synapses*. The study of brain function, therefore, must rest on a study of synaptic organization.

TRADITIONAL CONCEPTS OF
SYNAPTIC ORGANIZATION

The traditional view of the synaptic organization of neurons in the brain is illustrated in Fig. 1. We see here a sensory cell of the dorsal root ganglion (DRG), which receives stimuli in the periphery of the body and, through its long fiber, termed an *axon*, conveys signals to both the spinal cord and to higher centers. Within the spinal cord, the axon terminals make their specialized contacts —synapses—onto motoneurons, which are, by definition, the neurons that innervate the muscles. This is the classical reflex arc that was studied by Sherrington and that formed the basis of his great book, *The Integrative Action of the Nervous System* (1906).

The sensory fibers also connect to cells in higher centers (dorsal column nucleus), which, in turn, connect to cells in the thalamus. The thalamus is the great gateway to the neocortex of the cerebrum. From the cortex, pyramidal neurons send fibers to many regions, including those along the sensory input route (as shown in Fig. 1.). The longest fibers in the human brain reach all the way to the motoneurons in the spinal cord. In this manner are formed the complex loops and pathways that are the basis for delayed reflexes and conditioned responses, as well as the many kinds of motor, emotional, and intellectual behavior that originate within the organism itself.

Figure 1 illustrates that the brain consists of many local regions, or centers, and many pathways between them. At each center, the arriving axons make synapses onto the cell body (*soma*), and/or the short branched processes emanating from the cell body (*dendrites*), of the cells contained therein. Some of these cells send out a long axon that, in turn, carries the signals to other centers. These are termed *principal, relay,* or *projection* neurons. Other cells are concerned only with local processing within the center. These are termed *intrinsic* neurons, *local* neurons, or *inter*neurons. An example of this latter type is shown in the cerebral cortex in Fig. 1. The distinction between a principal and an intrinsic neuron

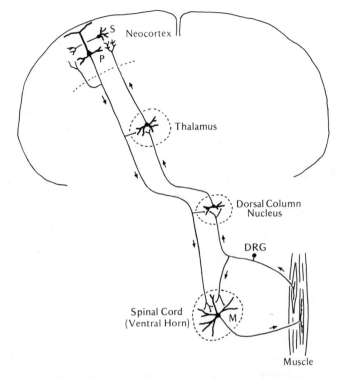

FIG. I. Examples of local regions and some circuits formed through the long axons of principal neurons. M, motoneuron; DRG, dorsal root ganglion cell; P, pyramidal (principal) neuron; S, stellate (intrinsic) neuron.

cannot be rigid, since principal neurons also take part in local interactions. It is, nonetheless, a useful way of characterizing nerve cells, which we will use throughout this book.

The principal and intrinsic neurons, together with the incoming input fibers, are the three types of neuronal constituents common to most regions of the brain. We will refer to them as a *triad* of synaptic elements. The relations between the three elements vary in different regions of the brain, and these variations underlie functional operations in processing inputs and relaying information to other centers of the brain.

Figure 1 illustrates the fact that each type of neuron is distinctive; neurons in different centers are of different size, form, extent, and branching pattern. The varieties of neurons, and their probable relationships to each other, were first revealed by the histologists of the 1880's. Their principal method was the Golgi stain, which has the remarkable property of staining only a few neurons in any given region, but staining them in their entirety. Figure 1 is, in fact, a composite drawing of Golgi-stained neurons. Chief among the classical histologists was Ramon y Cajal, who may be regarded as the first architect of the nervous system. His great textbook *Histologie du Système Nerveux* (1911) remains the most exhaustive and cogent account of the neural elements and circuits of the brain, as studied with the light microscope.

From Cajal and Sherrington and their contemporaries are derived the concepts of nervous organization that we regard as classical. Chief among them are the following. The circuits of the brain are built up of individual neurons. Each neuron has three parts: cell body, axon, and dendrites. Neurons transmit to each other through synapses. Transmission at synapses is one way; hence, the concept of "dynamic polarization" (Cajal) of the neuron, the neuron receiving synaptic inputs at its soma and dendrites and sending impulses out through its axon. The axons diverge (by branching) and converge (by overlapping onto a single target neuron). A neuron "integrates" its various inputs; the motoneuron, for example, acts as the "final common path" (Sherrington) for synaptic inputs controlling the muscles.

Over the intervening years, these concepts have appeared to be confirmed and extended by the anatomical and physiological studies of the long-distance pathways in the brain. Although never elevated to the status of a "central dogma" (wisely, as we shall see), they nonetheless have been widely accepted as the functional framework of the neuron doctrine. As such, they have been the basis of thinking not only in neuroanatomy and neurophysiology but also in psychology and the behavioral sciences, the clinical neurological sciences, and neural modeling, to say nothing of countless popular expositions of brain function.

SYNAPTIC ORGANIZATION OF LOCAL REGIONS

This body of principles was a remarkable achievement, particularly in view of the fact that it was based largely on deductions from the shapes of neurons and that proof for synaptic connections was lacking. The principles were derived almost exclusively from studies of principal relay neurons, with their obvious fiber connections to distant regions. Even in his original article on the neuron, Waldeyer (1891) explicitly pointed out that little could be said about the functional significance of dendrites or of neurons whose axonal arborizations were entirely local. Thus, it is probably fair to say that, from the start, the organization of local brain regions has never been adequately included within the framework of functional concepts built up around the neuron doctrine.

With the passage of time, progress in understanding the organization of local brain regions was very slow. For many regions, little could be said about the sites of synapses made by long axons, whether onto the principal neurons or the intrinsic neurons contained therein. Within a region, one could only speculate about the connections made by the intrinsic neurons. In most regions, there was little direct evidence about the physiological actions of particular synapses. The notion of inhibition was absent from the thinking of Cajal and of many who followed him. Finally, consistent definitions of an axon and a dendrite could not be agreed upon in the face of the enormous variety of processes exhibited by different neurons.

To all this was added the problem that local regions appeared to be distinctly different. Different types of neurons were present in each region, and specialized methods were necessary for studying them (i.e., visual stimuli for studying the visual regions, auditory stimuli for studying the auditory regions). This made it all the more difficult to perceive similarities in the connections and properties of neurons, which would provide a basis for principles of organization common to all regions.

These matters awaited evidence that could only be gained by improved methods, specifically, the use of microelectrodes, which can record from single, clearly identified nerve cells, and the electron microscope, which can reveal internal fine structure and give proof of synaptic connections. Both methods were introduced to the study of the brain in the 1950's, but the data they provided were not closely correlated until the 1960's. It is thus only in very recent years that results bearing directly on the long-unanswered questions about the synaptic organization of neurons in local regions of the brain have become available.

These results have been consistent with some of the classical notions, but inconsistent with others. Most unexpected have been the findings of synapses from axon to axon and from dendrite to dendrite, of a wide variety of functional properties of dendrites, and of synaptic interactions that take place without the mediation of nervous impulses. For most students and workers in the field, the brain was already complicated enough; to assimilate these new complexities was to be asked to pile Ossa on Pelion.

In sifting the new evidence two points have begun to emerge to provide a basis for a rational synthesis. One is that the concept of the single neuron as a functional unit—receiving information by way of its dendrites and sending information out by way of its axon—is much too limited and, in many cases, inaccurate. The new findings, particularly with regard to dendrites, indicate that a single neuron may contain many functional units in terms of its individual synaptic input-output relations and its dendritic compartmentalization. Also, any given neuron is only one small fragment of larger functional units made up of multineuronal assemblies. These considerations have indicated that a redefinition of the basic units of function in the nervous system is needed, and the suggestion has been made that, in its broadest sense, functional units may be defined as the morphological substrates for specific functions (Shepherd, 1972b). We will have ample occasion in the course of this book to observe why a definition of this generality is necessary to cover the types of organization seen in different regions of the mammalian brain.

The other point that has begun to emerge is that the local regions are perhaps not so different from each other as formerly appeared to be the case. At the synaptic level, particularly, similarities in structural patterns and functional properties have been found amid the welter of detail. The importance of identifying these similarities is that it is only against the background of what is common to different regions that the significance of their differences in structure and function can be assessed. This is no more than a truism that applies to all fields of scientific inquiry. At the least, it provides a starting point for the student in his initial approach to the subject; ultimately, it must provide the basis for deriving the principles of brain organization.

In this book, we will attempt a systematic comparison of those parts of the mammalian brain that have been most thoroughly studied. Our approach is therefore one of *comparative neuronal systems*. It is also *multidisciplinary* in that it draws on the coordinated evidence from anatomy, physiology, and biophysics. It focuses on the *structure-function relations* of synapses, since it is in the different kinds of junctional organization between neurons that our knowledge has undergone its most rapid increase, and this is where understanding must start, if one is to build up principles of organization at higher levels.

PLAN OF THE BOOK

We will begin by briefly reviewing the salient features of the structure of neurons and synapses and their functional properties. To understand the relation between structure and function, one must know the dynamics of electrical current flow in the neuronal processes. Because of the complexities of dendritic branching, this has been one of the greatest problems in the analysis of synaptic interactions. The solution to this problem, largely due to the work of Wilfrid Rall, has therefore been one of the most important developments of recent years. Rall's methods are mathematical and biophysical in character and have not been readily available to a wide range of students. We will therefore discuss

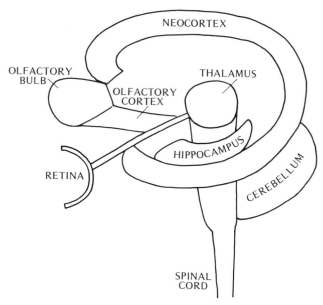

FIG. 2. Local regions for which synaptic organization will be described.

the essentials of these methods, in order to provide the basic tools for understanding the spread of activity in the complex branching trees of neuronal dendrites.

Figure 2 is a rough sketch which indicates the parts of the brain that will be studied in this book. Most of them are known even to the beginning student. The spinal motoneuron is first, as the classical subject for the study of central organization. The olfactory bulb is then introduced; its study has provided some of the clearest evidence about local synaptic organization, as well as support for concepts that go beyond the classical motoneuron model. The retina and cerebellum each have distinctive and stereotyped structures; together with the olfactory bulb, they are the best understood of the local regions of the brain. The thalamus has also received much attention; it forms a natural bridge to a consideration of the cerebral cortex. At this last level, it is now possible to consider all three types of cortex in the evolutionary scale: olfactory, hippocampal, and neocortical.

In each chapter, the same sequence will be followed: First, a description of the main types of neuronal elements; second, a description of the major types of synaptic connections between the neurons; third, a summary of organization in terms of a basic circuit diagram, with a discussion of the cardinal aspects of the organization of that region; fourth, the dynamics of the local circuits in terms of the main types of synaptic actions; and fifth, a consideration of the dendritic properties crucial to the synaptic interactions. Keeping to this sequence will, it is hoped, make it easier for the reader to compare the different regions.

DIMENSIONS The dimensions of nerve cells and their processes are important factors in the study of synaptic organization. The classical histologists who studied Golgi-stained neurons were primarily interested in describing the marvelously varied shapes of neurons and the distributions of the processes, and they paid relatively little attention to dimensions. Cajal, for example, provided no scales for the diagrams in his monumental survey of neuronal architecture (Cajal, 1911). In modern studies of fine structure with the electron microscope, there has been a similar tendency to emphasize the shapes of processes and terminals and the patterns of synaptic connections. The application of methods for analyzing electrical current flows, however, requires precise figures for the dimensions of axonal and dendritic processes. Such figures are necessary, not only for understanding synaptic integration in neurons of a given region but also for comparing the organization of different regions.

A scale for lineal (distance) measurements is provided in Fig. 3. Just as we live our daily lives in a world of inches and feet (or centimeters and meters), so the student of synaptic organization lives in a world of micrometers and hundreds of micrometers. As shown in Fig. 3, a micrometer (μm) is one-millionth part of a meter, one ten-thousandth part of a centimeter, or one-thousandth part of a millimeter. Its colloquial name is "micron" and it will be the common currency of our descriptions in this book.

In comparing dimensions in different parts of the brain, the

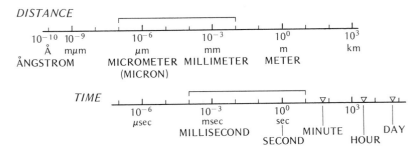

FIG. 3. Scales (logarithmic) for distance and time. Brackets, ranges relevant to the synaptic organization of local regions.

student is never more frustrated than by the multiplicity of scales and magnifications that so often confront him. We will try to ease at least this much of the burden by the use of only two scales in the later chapters. For the diagrams of neuronal elements, all magnifications will be ×50; i.e., 5 mm in the diagram equals 100 μm actual length. For diagrams of synaptic connections, all magnifications will be 100 times this, or ×5000; i.e., 5 mm in the diagram equals 1 μm actual length. By this systematic approach, some dramatic and important differences in size between processes of different regions will be brought out.

Figure 3 also provides a temporal (time) scale for measurements. What the micron is to the dimensions of synaptic organization, the millisecond (one-thousandth of a second; msec) is to the time course of synaptic actions. As it turns out, the time courses of synaptic actions are too varied to permit their representation on only one time scale, so the student must pay particular attention to the scales in comparing recordings of synaptic actions in different regions. The number of different time scales has, in any case, been held to a minimum to facilitate comparisons as much as possible.

The dimensions of space and time involve only the simplest concepts of quantitation, yet the student will find that their usefulness as tools in the analysis of synaptic organization requires a facility that comes only with practice.

Part 1

PRINCIPLES OF STRUCTURE AND FUNCTION

2

AXONS, DENDRITES, AND SYNAPSES

The study of synaptic organization requires, first, the identification of the structures that take part in synaptic connections, and we will, therefore, begin with an overview of the current knowledge of the parts of the neuron and the different kinds of synapses. The identification of neuronal parts turns out to be quite difficult because of the extraordinary variety of forms neurons and their processes assume. The types of neurons represent almost every imaginable extreme, and even within a type it is fair to say that no two neurons are exactly alike. In an attempt to make sense of this bewildering variety, the classical histologists divided neuronal processes into *axons* and *dendrites*, on the basis of external shape, extent, and branching pattern. This distinction is clear cut only in certain simple cases, however, and, ultimately, our definitions must be based on fine structural criteria. The brief description that follows may be supplemented by the accounts of De Robertis, Nowinski, and Saez (1970); Peters, Palay, and Webster (1970); and Bodian (1972).

PARTS OF THE NEURON

The nerve cell, like all other cells of the body, is bounded by a *plasma membrane*. In cross sections of electron micrographs at low magnification (as in the diagrams of fine structure in this

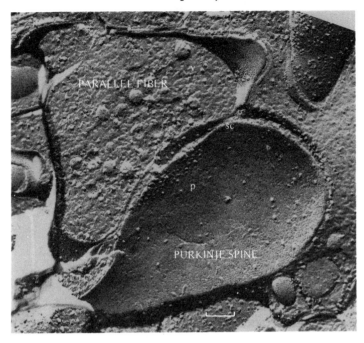

FIG. 4. A freeze-fractured specimen of synaptic terminals in the cerebellum. The line of cleavage is such that one sees the inner surface of an outer leaflet of a Purkinje cell dendritic spine; it also cuts across a synaptic terminal of a parallel fiber. Note the vesicles (v) in the presynaptic terminal, the widened synaptic cleft (sc), and the accumulation of small particles (p) in the postsynaptic terminal membrane. Bracket 0.1 μm. (From Landis and Reese, 1974.)

book) the membrane appears under the electron microscope as a single dark line, about 80 Å thick (i.e., just less than one one-hundredth of a micron). At higher magnifications it can be seen that there are, in fact, two dark lines with a light space between them. This gives the membrane a three-layered structure, with inner and outer leaflets.

The plasma membrane has been vividly revealed in recent studies employing the freeze-fracturing technique, in which a tiny block of tissue is frozen and then fractured by a swift blow with

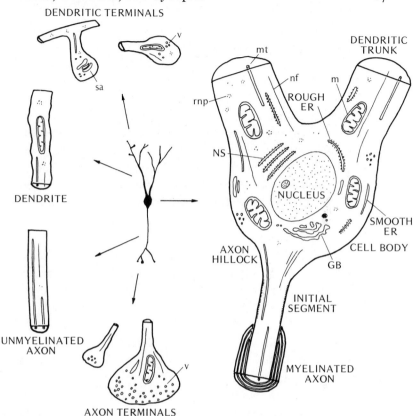

FIG. 5. Diagrams of the parts of the neuron. A principal neuron (as stained by the Golgi method or by intracellular injection of dyes), shown at center, is surrounded by schematic drawings of fine structure (as viewed in the electron microscope) of the different parts. ER, endoplasmic reticulum; GB, Golgi body; NS, Nissl substance; mt, microtubule; nf, neurofilament; rnp, ribonucleic particles; sa, spine apparatus; v, vesicles; m, mitochondria.

a sharp blade. A micrograph of a specimen prepared by this ingenious technique is shown in Fig. 4. The lines of cleavage are, unexpectedly, not between the membranes of the two neighboring neurons, but between the inner and outer leaflets of the same membrane. As can be seen in Fig. 4, the fracture line jumps from one membrane to the next, or cuts entirely through a process, in its course through the tissue. This technique is just one of several new procedures developed recently for studying the internal composition of axons and dendrites and the ultrastructure of membranes and synaptic junctions.

The three-layered structure comprises the *unit membrane,* common to all animal cells, which is thought to consist of oriented lipid and protein complexes. In all cells of the body this membrane controls the interchange of substances between the cell and its environment. In nerve cells it is, additionally, the site of origin of the electrical activity that is the basis of nervous signals. There is as yet no evidence for differences in membrane structure in different parts of the neuron. Hence, identification of neuronal parts must depend upon other criteria.

The fine structure of different parts of the neuron is summarized in Fig. 5, which shows a representative neuron surrounded by schematic drawings of electron micrographs of the various parts.

CELL BODY   The cell body (soma) by definition is the part of the neuron that contains the *nucleus* (see Fig. 5). As in all cells, the nucleus contains the genetic material of the neuron. The size of the cell body varies widely in different neurons, from 6 μm to 100 μm or more in diameter. The size of the nucleus similarly varies. It tends to fill nearly all the cell body of smaller cells, leaving only a narrow rim of cytoplasm. Larger neurons, on the other hand, tend to have relatively more cytoplasm.

The cytoplasm surrounding the nucleus, the *perikaryon,* contains a widespread system of membranes, the *endoplasmic reticulum* (ER), and numerous small particles, the *ribosomes.* The latter are the principal sites of protein synthesis in the cell. The distribution of the ribosomes is the basis for a fundamental distinction between ER that is covered with closely adhering ribosomes (granular, or rough, ER) and ER that has no adhering ribosomes (agranular, or smooth, ER).

In some parts of the perikaryon, the rough ER is organized into aggregations of pancake-like folds, or cisternae. These are called *Nissl bodies,* which are regarded as nodal points in the ER for the synthesis of structural and secretory proteins, including substances that play a role in synaptic transmission.

The smooth ER, on the other hand, is specialized in places to form stacks of cisternae and assorted vacuoles, the *Golgi apparatus,* implicated in the turnover of vesicle membranes and the packaging of secretory substances involved in synaptic transmission.

Mitochondria are scattered throughout the perikaryon. In vertebrate neurons, they range from 0.1–0.5 μm in diameter and up to several microns in length. They are often found in the vicinity of Nissl bodies, where, as in other cells of the body, they supply energy for metabolism.

Two elongated structures within the perikaryon are *microtubules* and *neurofilaments.* Microtubules are 200–300 Å in diameter, whereas neurofilaments are about 100 Å in diameter. At high resolution, neurofilaments have a light core, hence, they also appear as tubules. Both these elements, singly or in aggregates, weave their way through the perikaryon. Microtubules are composed of protein and are thought to be involved in several functions, including cell division, the shaping of cell processes during development, and intraneuronal transport of substances; they may also have contractile properties. Neurofilaments are also thought to be protein. Their function is enigmatic, beyond the obvious one of imparting some structural rigidity; one possibility is that they may constitute a stage in the turnover of microtubule material.

Vesicles should also be included among the constituents of the perikaryon. They may be found singly here and there or in small clumps near the border. In the latter case, they form part of the presynaptic apparatus of a synapse from the cell body to a neighboring neuron. They will be described in detail later in this chapter.

Other types of organelles within neurons include subsurface cisterna; various types of large vesicles, including lysozymes; multivesicular bodies; and various inclusion bodies.

All the main internal structures of the perikaryon described above are common to animal cells, in general, with the exception of neurofilaments. An interesting finding is that the distribution of internal structures appears to be distinctive for some types of neurons. This suggestion that neurons with distinct geometrical shapes may have distinct patterns of internal organization within their perikarya is well worth further study. The significance of the differences between neurons is not yet clear; there is no correlation, for instance, between Nissl body occurrence and neuronal size.

AXON HILLOCK   We can now begin to consider the types of processes that arise from the cell body. In neurons that have an axon, the axon is often seen to arise from a cone-shaped region of the perikaryon or a large dendrite, the *axon hillock*. As shown in Fig. 5, the hillock is characterized by decreased density of ribosomes and other organelles and funneling of microtubules and neurofilaments; the former are typically grouped in fascicles. An axon hillock is particularly characteristic of large neurons with large axons, but it is by no means clear that an axon hillock, as characterized above, is present in all neurons, and especially in the smallest neurons with the finest axons.

INITIAL SEGMENT   In neurons that give rise to a myelinated axon, the *initial segment* is that portion between the axon hillock and the point at which myelination begins. As indicated in Fig. 5, it is characterized by a thin layer of dense material underneath the plasma membrane, the so-called dense undercoating. Microtubules continue, singly or in fascicles, from the axon hillock into the initial segment. A few ribosomes may be present, but the other organelles of the perikaryon are absent. The initial segment is characteristic of neurons giving rise to myelinated axons, although it is absent in the Purkinje cell of the cerebellum (Chapter 8). It is also doubtful whether the initial segment can be considered to be present, as an entity, in the smallest neurons giving rise to unmyelinated axons.

AXON   The long single nerve process that extends from the axon hillock and initial segment is, by definition, an axon. The axon, thus, can always be identified if it has this origin. The fine structure of the axon is characterized by neurofilaments, microtubules, mitochondria, and smooth ER; generally, ribosomes are few or absent. The relative number of these structures depends on size. Large axons (up to 20 μm in diameter in the vertebrate brain) tend to contain many neurofilaments but relatively few microtubules. In small axons, however, the ratio is reversed; in the thinnest axons (0.1 μm) only a few microtubules may be present.

The criteria for identifying an axon include the sheaths that surround them. In their course through either the periphery or the brain, axons are enveloped by the processes of satellite cells, called Schwann cells. The Schwann cells enwrap the larger axons in many layers of plasma membrane; these tightly layered membranes are called *myelin.* In general, the larger the axon, the thicker the myelin, up to a hundred layers or so around the largest axons. Along its length an axon is enwrapped by a series of Schwann cells. The place where two Schwann cells meet is free of myelin for a micron or so, and this region is called a *node of Ranvier.* At the node, there is an undercoating of the plasma membrane similar to that of the initial segment. The internodal distance is characteristically 1 mm or so. As we shall see, myelin is a specialization that provides for rapid impulse transmission from node to node.

Although myelination is the most important distinguishing characteristic of large axons, the finest axons (below about 1 μm diameter) are unmyelinated. Since fine axons predominate in the brain, both in the pathways between regions and within regions, other criteria for axon identification are needed when an axon is viewed at a distance from its cell body. As we shall see, it is difficult, in electron micrographs of the neuropil of local regions, to distinguish fine axons from fine dendtritic processes.

AXONAL BRANCHES   Axons characteristically have branches that distribute the signals traveling in them to more than one destination. The branches take several forms. An axon may simply divide into two branches of equal diameter; an example is the axon of the granule cell of the cerebellum (Chapter 8). An axon may give off a relatively large branch, referred to as a *collateral*. Such a branch may arise during the long-distance course of the axon, as, for example, the collateral from the axon of a DRG cell to the spinal cord (see Fig. 1). Or a collateral may arise locally within the region of origin. Such a branch may have a generally lateral orientation, or it may run backwards within the region, in which case it is called a *recurrent collateral*; an example is shown in the cortical pyramidal cell in Fig. 1. We will find that recurrent collaterals are important components in the synaptic circuits of local regions.

It is often stated that axons maintain their diameters during their course, in contrast to dendrites, which typically taper—and that axonal branches arise at right angles from their parent fibers, whereas dendritic branches arise at acute angles. Although these features are found in many neurons, they tend to be associated with neurons in which axon and dendrite are most differentiated, the one from the other; they are unfortunately less useful in identifying neuronal processes that are not so clearly differentiated.

The fine structure of axonal branches is generally similar to that of the parent fiber, being characterized mainly by variable numbers of microtubules. In axons that give rise to synapses as they pass by their target neurons (*en passage*), groups of vesicles are present in relation to synaptic sites; an example is the parallel fiber of the cerebellum.

DENDRITES   We have already mentioned some of the ambiguities of identification of neuronal parts. The problem becomes most difficult with regard to dendrites.

The simplest case we can consider is that of a principal neuron (note again the several examples in Fig. 1). All principal neurons have, by definition, a long axon. In these neurons, any process that is not the axon arising from the axon hillock and initial segment may be defined as a dendrite. The stout trunks of dendrites have much the same internal structure as that of the cell body from which they arise. Deiter's original term "protoplasmic prolongations" (1865) can, therefore, be said to have been right on the mark. Microtubules are prominent in the dendritic trunks, often present in orderly arrays, whereas neurofilaments are few; this is the reverse of the ratio for large axons. As in the soma, vesicles may be found, either scattered or in presynaptic aggregations.

This simple definition of dendrite, as a process that is not an axon, would appear to be general enough, but, unfortunately, it runs afoul the fantastic variety of neuronal processes. It does not even apply to all principal neurons. In invertebrates, for example, the cell body characteristically gives rise to a single process, which becomes the axon, and which also gives off a number of branches to the local region before projecting to distant regions. Our present knowledge is not sufficient to enable us to conclude whether the single process and the local branches are axonal or dendritic, and the usual practice is therefore to refer to them noncommittally as "processes" or "neurites".

In the vertebrate brain, the DRG cell (see Fig. 1) is a variety of principal neuron that has become specialized in the mammal for long-distance transmission between the periphery and the spinal cord. In this specialization, its central "axon" and peripheral "dendrite" have fused into one long fiber. The fact that the peripheral process is embryologically a "dendrite" but structurally an "axon" has presented a terminological dilemma unresolved to the present day.

The problem is even more difficult for intrinsic neurons; they are involved exclusively with local interactions and, therefore, do not require an axon for long-distance transmission. As a result, some intrinsic neurons have a morphologically identifiable axon, whereas others do not. In the latter case the definition of a dendrite as any process that is not the axon of a particular cell will not do, since the particular cell does not have an axon to begin with. The granule cell of the olfactory bulb (Chapter 6) and the amacrine cell of the retina (Chapter 8) are examples of this type of "anaxonal" neuron. In such cells, the common-sense approach is to call these processes dendrites on the basis of their similarity to dendrites of other neurons or else to use the noncommittal term "process."

These are only a few illustrations of the problem attending the definition of dendrite, as distinct from axon. They by no means exhaust the list. A further complication, for example, is the fact that myelination is not exclusively associated with axons; myelin is also found around cell bodies and dendrites in certain instances, as we shall see in the olfactory bulb. A myelinated structure, therefore, cannot be said to be, by definition, an axon; conversely, an unmyelinated structure may be either an axon or a dendrite.

It is often stated that the presence of ribosomes is a distinguishing characteristic of dendrites, but it is becoming clear that this generalization has limited usefulness. It has yet to be established that ribosomes are lacking in all the short unmyelinated axons of the brain. On the other hand, many processes that, by other criteria, would be accepted as dendritic contain few or no ribosomes.

As one proceeds from the dendritic trunk, through the larger to the smaller dendritic branches, one encounters occasional membranous cisternae and variable numbers of microtubules. A small branch may have lighter- or darker-staining background material and may contain a few ribosomes or none at all. Vesicles may be scattered about or found in presynaptic groups. In the smaller branches mitochondria become especially prominent, usually slender, and sometimes much elongated; their lengths may reach as far as 20 μm. By the time one reaches the smaller dendritic branches, the internal structure is not significantly different from that of axons and axonal branches of comparable size. In some cases, a more irregular contour and a more variable orientation aid in identification. But both the axonal and non-axonal (i.e., dendritic) processes of nerve cells are too variable to admit of generalization with regard to shape and extent.

TERMINALS After dividing into smaller branches, axons end in terminals of various descriptions, from simple enlargements (but-

tons, boutons, knobs, end-feet) to elaborate claws, mossy terminals, and other complicated configurations. Dendritic branches
also terminate in a variety of ways, as simple enlargements (varicosities, knobs, spines, gemmules), as cilia or other fine elongated
processes, as claws, and even as sheets. Apart from these specific
shapes assumed by different terminals, no general statement can
be made about their internal fine structure. Of the organelles we
have considered, ER and ribosomes are usually scarce, mitochondria are usually plentiful, and microtubules and neurofilaments
are variably present. Vesicles are usually (but not always) the
most conspicuous component of terminals; they are either scattered or accumulated in dense clouds within the terminal cytoplasm. In addition, vesicles are organized in smaller groups in relation to synaptic contacts, as will be described in the next section.

Traditionally, electron microscopists have used the presence or
absence of synaptic vesicles as the criterion for distinguishing an
axon from a dendritic terminal when identifying profiles seen in
electron micrographs; according to the functional interpretations
of the neuron doctrine, axon terminals contain synaptic vesicles
and dendritic terminals do not. The discovery of presynaptic
dendrites, however, has rendered this criterion untenable. *A terminal cannot be identified as axonal solely on the basis of the fact
that it contains synaptic vesicles.* In some cases, a relatively high
density of vesicles is characteristic of axonal terminals; in others,
the presence of such a distinguishing organelle as a spine apparatus (an aggregation of cisternae) indicates a dendritic spine. But,
in general, we must conclude that, on the basis of fine structure,
the terminals of axons and dendrites form a single, very broad
class. It may be noted that this is not at all inconsistent with the
classical term "telodendria" for the terminals of axons.

A point perhaps worth noting is that the term "terminal",
strictly interpreted, refers to the geometry of a process, i.e., the
end of a branch or an excrescence from a branch. Presynaptic loci
characteristically occur in geometrical terminals, but enough has
been said here to indicate that they may also occur along the
branches of an axon, along the branches and trunks of dendrites,

and at the cell body. Terminal as an appellation for any presynaptic locus is firmly fixed in the literature, however, and we will use it in this more general sense, as well as in the stricter geometrical sense.

TYPES OF SYNAPSES

In reviewing the fine structure of neuronal parts, we have avoided considering, as far as possible, the organelles that are related specifically to synapses. The reason for this has already been indicated, that anatomists in the past, following classical concepts, have used the orientation of synapses as the basis for distinguishing between axon and dendrite and have been led thereby into error. Let us therefore consider synaptic morphology independently of neuronal morphology and then bring the two together.

MEMBRANE JUXTAPOSITIONS   There are several types of structural relations one neuron can have with another. The simplest is a juxtaposition of their membranes, the two being separated by the ubiquitous extracellular space (cleft) of about 200 Å. This is illustrated schematically in Fig. 6. Some of the fine unmyelinated axons (e.g., olfactory axons, parallel fibers of the cerebellum) have this membrane-to-membrane relationship with each other. It also occurs throughout the neuropil of the local regions of the brain, between dendritic processes or between closely packed cell bodies, as well as between neurons and their satellite cells. Juxtaposition of membranes provides for several possible functions, including ionic or metabolic effects, mediated by extrusion, and uptake of substances into and out of the intervening extracellular cleft. It may also provide for electrical interactions between neurons under some conditions; the site at which such an interaction occurs is an *ephapse*.

MEMBRANE APPOSITIONS   The next stage of relatedness between neurons is a specific apposition of their membranes. This occurs at sites where (1) the two membranes come close together or are

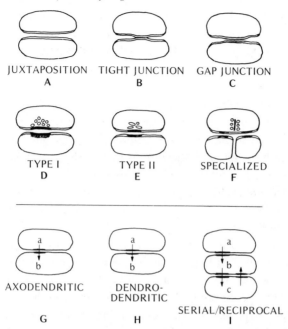

FIG. 6. Types of synapses. (A) Juxtaposition (non-synaptic) of membranes; (B) tight junction; (C) gap junction; (D) type I chemical synapse; (E) type II chemical synapse; (F) specialized terminal with specialized synapse.

Types of synaptic arrangements. (G) Axodendritic, from axon (a) to dendrite (d); (H) dendrodendritic, from dendrite (a) to dendrite (b); (I) serial synapse from (a) to (b) to (c) and reciprocal synapses from (b) to (c) and from (c) to (b).

fused and/or (2) the membranes appear more dense. Such sites are found between cells throughout the body. Depending on details of structure, they are called occluding junctions, desmosomes, tight junctions, gap junctions, septate junctions, zonulae adherens, etc. (See Peters et al., 1970). They vary widely in size and form, ranging from small spots to long strips or patches. Such junctions provide for several possible functions: simple adhesion; transfer of substances during metabolism or embryological development; restriction of movement of substances in the extracellular compartment. An instance of the latter function is provided by the *tight junctions* between the cells that line the blood vessels and ventricles of the brain. The two outer leaflets of the unit membrane of these junctions, as illustrated in Fig. 6, are completely fused, to form a five-layered complex. These tight junctions restrict the movement of substances in the extracellular space and are responsible for the so-called blood-brain barrier.

An important type of membrane apposition in the nervous system is the so-called *gap junction*. Here, the outer leaflets are separated by a gap of 20-40 μm, to form a seven-layered complex (see Fig. 6). In several cases, the presence of these junctions has been correlated with the physiological finding of a low-resistance electrical pathway between two neurons (see Reese and Brightman, 1969; Bennett, 1972). On this basis they have been categorized as *electrical synapses*. The junction varies in diameter from 0.1-10 μm. At high resolution, dense material is seen beneath each apposed membrane, and it can be shown that the membranes are part of two systems of channels, the one continuous with the extracellular space, the other connecting the two cells. Electrical synapses are a common form of interneuronal connection in lower vertebrates (Bennett, 1972); they have been less frequently described in higher vertebrates, but our knowledge is expanding rapidly in this area. Gap junctions are found between many types of cells in the body besides neurons. In addition to electrical coupling, their possible functions include those mentioned above for closely apposed membranes.

CHEMICAL SYNAPSES The most complicated type of junction in the nervous system, and the type considered to be the most characteristic, is the chemical synapse. It differs morphologically from other types of membrane appositions in being strongly oriented, or polarized, from one neuron to the other. This polarization is determined mainly by two features: (1) an unequal *densification* of the two apposed membranes and (2) the presence of a group of small *vesicles* near the synaptic site. In certain cases (e.g., the neuromuscular junction), it can be shown unequivocally that transmission is from the vesicle-containing process to the other process, and one has, therefore, the terminology of a *presynaptic* process and a *postsynaptic* process, respectively.

At an early stage in studies with the electron microscope, Gray (1959), working on the cerebral cortex, obtained evidence that synapses belong to two morphological types. There is a growing consensus that, despite many local variations and gradations between the two, this division has some validity. The two types are illustrated in Fig. 6. The distinguishing features may be summarized as follows. Type I: synaptic cleft approximately 300 Å; junctional area relatively large (up to 1-2 μm in extent); prominent accumulation of dense material next to the postsynaptic membrane (i.e., an asymmetric densification of the two apposed membranes). Type II: synaptic cleft approximately 200 Å; junctional area relatively small (less than 1 μm in extent); membrane densifications modest and symmetrical.

Following the recognition of these types, evidence was obtained (see Uchizono, 1965) that, in many parts of the brain, type I synapses are associated with large spherical vesicles (diameter approximately 300-600 Å) which are usually present in considerable numbers. Type II synapses, on the other hand, are associated with smaller (100-300 Å diameter) vesicles, which are less numerous and which, significantly, take on various ellipsoidal and flattened shapes. The distinction between round and flat types of vesicles is by no means a sharp one; in many synapses a vesicle simply tends to the one shape or the other.

These two morphological features (symmetry of membrane density and shape of vesicles) have provided a convenient means for characterizing synapses, and we will use the terms type I and type II in this sense to describe synapses in different regions of the brain.

The recognition of these two types of synapse has provided anatomists with a most useful tool to unravel the synaptic organization of local brain regions. Much of this usefulness has been based on the premise that all the synapses made by a given neuron onto other neurons are either of one type or the other. This is commonly called the *morphological corollary of Dale's Law,*

Dale's Law being usually understood as stating that a given neuron has the same physiological action at all its synapses. As we will see in the next chapter, this is neither what Dale, in fact, put forward nor what electrophysiology reveals. Nor has it been proved that the morphological corollary has universal validity; in the cochlear nucleus, for example, there is evidence that an auditory axon makes both types of synapses onto the same postsynaptic neuron (Kane, 1973). In the regions of the brain under review here, the premise that all the synapses made by a given neuron are of the same type has been generally assumed and thus far substantiated.

Many neuroanatomists have been skeptical of the validity of the two types of synapse on the basis of the fact that the flattening of vesicles has been shown to depend on the osmolarity of solutions used in preparing the tissue for electron microscopy (Valdivia, 1971). But, in a sense, everything the electron microscopist sees is a distortion of the true dynamic living state. The interpretation of electron micrographs, and of any preparations of anatomical specimens, must be made with this constantly in mind. That the recognition of the two types of synapse has been the basis for remarkable progress in the understanding of the synaptic organization of the brain may be regarded as sufficient reason for using it as the basis for our review of present knowledge.

These two types of synapse provide relatively small areas of contact between neurons. They may be characterized as *simple* synapses. They are typical of the contacts made by small terminals, both axonal and dendritic, and they are also the type of contact made by most cell bodies and dendrites when those structures occupy presynaptic positions. It is probably fair to say that they make up the majority of synapses in the brain. This, in itself, bespeaks an important principle of brain organization, that the output of a neuron is fractionated, as it were, through many synapses onto many other neurons and, conversely, that synapses from many sources play onto a given neuron. This is an essential aspect of the complexity of information processing in the brain.

In addition, there are, in many regions, much more extensive contacts with more elaborate structure that may be characterized as *specialized synapses*. The neuromuscular junction is an example in the peripheral nervous system (Chapter 3). In the central nervous system, we find an example in the retina, where the large terminal of a receptor cell makes contact with several postsynaptic neurons; within the terminal, the synaptic vesicles are grouped around a special small dense bar. This arrangement is shown very schematically in Fig. 6 (F) and is described in detail in Chapter 7.

One may also characterize the terminal structures in the geometrical sense, as previously defined. A terminal may be small and have a single synapse onto a single postsynaptic structure, as shown in most of the diagrams of Fig. 6. These may be characterized as *simple terminals*. On the other hand, a large terminal, with complicated geometry, may be characterized as a *specialized terminal:* examples are the neuromuscular junction (Chapter 3) and the basket cell endings around the Purkinje cell (Chapter 8). In many regions of the brain, large terminals have synapses onto more than one postsynaptic structure; the receptor terminal in the retina mentioned above is an example. Another example is the large terminal rosette of the mossy fiber in the cerebellum, which has as many as 300 synaptic contacts onto postsynaptic structures.

Within the brain are all possible combinations of synapses and terminals. Simple synapses may be established by any of the parts of the neuron: terminals, trunks, or the cell body. Simple synapses may also be made by specialized terminals, as in the case of the mossy fiber of the cerebellum (Chapter 8). On the other hand, specialized synapses may be made by small terminals, as in the spinule synapses of the hippocampus (Chapter 11), and, finally, specialized synapses may arise from specialized terminals, as in the case of the retinal receptor.

PATTERNS OF SYNAPTIC CONNECTIONS Synapses are also categorized by the kinds of processes that take part in the synapse. Thus, for example, a contact from an axon onto a cell body is termed an *axosomatic* synapse, whereas that onto a dendrite is termed an *axodendritic* synapse [Fig. 6 (F)]. Similarly, a contact between two axons is termed an *axoaxonic* synapse, and a contact between two dendrites is termed a *dendrodendritic* synapse [Fig. 6 (G)]. In the diagrams of Fig. 6 (A)-(E), either one of the processes can be regarded as axonal or dendritic to represent these simple arrangements.

A single synapse seldom occurs in isolation in the brain; it is usually one of a number of synapses that together make up a larger pattern of interconnecting synapses. The simplest of these patterns is that formed by two or more synapses situated near each other and oriented in the same direction, i.e., they are all axondendritic. A more complicated pattern is one in which there is a synapse from process (a) to process (b), and another from (b) to (c). Such a situation is diagrammed in Fig. 6 (I). These are referred to as *serial* synapses; examples are axoaxodendritic sequences and axodendrodendritic sequences.

Another pattern has a synapse from process (b) to process (c), and a return synapse from (c) to (b). This is also diagrammed in Fig. 6 (I). It is referred to as a *reciprocal* synapse. If the two synapses are side by side, they are called a *reciprocal pair*. The dendrodendritic synapses between mitral and granule cells in the olfactory bulb are of this type. If the two synapses are far apart, a *reciprocal arrangement* results; there are several examples of this type (Chapters 6, 7, and 9).

The first synapses identified by electron microscopists were simple contacts made by simple terminals, of the axosomatic and axodendritic type. Since these simple arrangements were in accord with the functional concepts of the neuron doctrine, as summarized in Chapter 1, they came to be regarded as "classical" synapses. The axoaxonic and dendrodendritic types were identified later, as were the serial and reciprocal arrangements and the various types of specialized synaptic contacts and terminals. Since these synapses, terminals, and patterns did not fit classical concepts, the practice grew up of referring to the simple synapses as "conventional" and to all the other synapses as "unconventional" or even "nonusual."

In the brain, as in society, such terms carry inevitable overtones of moral opprobrium, and it is wisest to avoid them. Suffice it to say that there is probably no more certain sign of the obsolescence of an idea than the practice of labeling as "unconventional" those facts that do not fit it. The nervous system does not put these labels on its synapses. We may conceive that, in any given region, it is faced with specific tasks of information processing, and it assembles the necessary circuits according to those functional demands from the available neuronal components. That many of the ways by which these components are connected do not fit classical concepts is good reason to revise these concepts. We will be taking a first step in that direction if we characterize synapses in terms of their relative complexity and specific patterns of connection, as we have attempted to do in the discussion above.

VESICLES Synaptic vesicles are a subject in themselves. They come, in the felicitous phraseology of Palay (1967), like chocolates, in a variety of shapes and sizes, and are stuffed with different kinds of filling. Small vesicles (200-400 Å in diameter) are the most common; they are the ones we have discussed in regard to type I and type II synapses. At some of these synapses, there is evidence that acetylcholine is bound to, or contained within, the vesicles; such synapses are, therefore, called cholinergic. At other synapses, the vesicles appear to be associated with certain amino acids. These are the putative transmitter substances that are released by the presynaptic terminal when it is activated and that mediate the synaptic action onto the postsynaptic membrane, as will be described in the next chapter.

Another type of vesicle is medium sized (500-900 Å in diameter) and contains a dense granule; these vesicles are associated with the transmitter norepinephrine. Their synapses are of the adrenergic type found in the autonomic nervous system and in certain parts of the brain. Large vesicles (1200-1500 Å in diameter) are characteristically found in neurosecretory cells, for example, in the nerve endings of hypothalamic neurons, which send their axons to the pituitary. A large dense droplet within these vesicles contains a polypeptide hormone, which is released in response to the appropriate behavioral stimulus.

This very brief account only scratches the surface of the subject of synaptic vesicles, and the reader is referred to Peters et al. (1970) for further details. The main point to be made here is that the vesicles are structural evidence of the fact that chemical synapses constitute a variety of neurosecretory apparatus. Like other secretory mechanisms, they are activated by specific stimuli, have specific targets, and exert particular actions on those targets. The dynamics of these mechanisms are the subject of the next chapter.

SUMMARY

It may be concluded that the division of neuronal processes into axon and dendrite can be made only for the largest processes of principal neurons and certain intrinsic neurons. The smaller processes of all types of neuron overlap too much in their morphology to allow simple generalization. Similarly, the synaptic connections between neurons vary greatly. Although small contacts made by single small terminals onto single postsynaptic structures are the major type over all, in any given region, many if not most of the contacts may be highly specialized in terms of type of contact, type of terminal, and patterns of interconnection.

In the practical problem of analyzing synaptic organization, the electron microscopist, confronted with a single view of a central region, must seek supplementary evidence in order to differentiate between the processes and to identify the synapses and the cell types from which they arise. This evidence is provided by serial reconstructions of many sections to give a three-dimensional view; by close correlations with Golgi-stained material; by study of neurons that have been stained intracellularly or by the Golgi method; and by observations of neurons undergoing selective degeneration. These methods have been developed and increasingly used in recent years, and have provided the evidence for the account of synaptic organization presented here.

3

ACTION POTENTIALS
AND
SYNAPTIC POTENTIALS

It is characteristic of neurons that they not only have different parts, as we have seen, but also different functional properties. One of the chief goals of physiological analysis is to determine the properties of the different neuronal parts. This is the necessary basis for reconstructing the dynamics of synaptic organization.

If neurons are remarkable for their variety of forms, they are no less remarkable for their variety of functions. We must begin, therefore, by recognizing the extensive range of these functions; it includes those related to embryological growth and differentiation, metabolism, response to injury or denervation, internal movement of substances, humoral secretion and reception, and long-term changes in excitability, to name only the most obvious. Our concern must be restricted to those aspects of information transfer that involve relatively rapid electrical and chemical changes. These are the basis for the ongoing activity and behavior of all forms of animal life, and they are also an essential part of the fabric of other functions.

MEMBRANE POTENTIAL

It is customary to represent functional properties of the neuronal membrane in terms of a simple model for a single locus. We con-

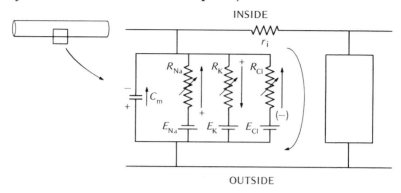

FIG. 7. Equivalent electrical circuit for a patch of neuronal membrane. C_m, capacitance; R (resistance) divided into channels for Na (sodium), K (potassium), and Cl (chloride) ions; E, driving forces for ionic movements. Inside arrows, direction of ion flows down electrochemical gradients; outside arrows, direction of net current flow in membrane capacitance of this and the neighboring locus, for the case of a Na conductance change.

sider therefore a locus, or "patch", of neuronal membrane and represent its electrical properties with a simple circuit, as in Fig. 7. In this circuit, there is a capacitance (C_m), due to the lipids in the membrane. There is also an electrical resistance (R), which is divided into conductance channels (Conductance $=$ 1/Resistance) for the flow of ions. In series with the ionic conductances are batteries representing sources of electrochemical potential, which serve as the driving forces to move the various ions through their channels. We consider two such batteries, oppositely directed, which move their positively charged ions in opposite directions across the membrane. The inward-directed battery (E_{Na}) will represent the force that drives sodium ions (Na^+) through their conductance channel (R_{Na}). The outward-directed battery (E_K) will represent the force that drives potassium ions (K^+) through their conductance channel (R_K).

The driving forces for these ions derive from the differences in concentrations of ions across the membrane, there being, as in all cells of the body, a relatively high concentration of Na outside the cell and a relatively high concentration of K inside. The batteries, therefore, represent the tendencies of the ions to move passively down their concentration gradients. The concentration differences are maintained in the resting cell by relatively low permeabilities to ion movement across the membrane and by a metabolic "pump" that takes in K and extrudes Na. The pump requires ATP for its energy and is the functional link between the electrical properties of the membrane and the metabolic system of the cell.

Due to a selective permeability to K ion in the resting membrane, there is an excess of total positive charge on the outside of the membrane capacitance, and an excess of negative charge on the inside. This creates the resting *membrane potential* (V_m); again, it is a characteristic of all cells. The dynamics of synaptic organization are largely concerned with moment-to-moment competition for control of the membrane potential in different parts of the neuron.

It is traditional to think of the membrane potential as having a value of some -70 to -90 mV (inside negative relative to outside), as it is in many cells of the body, in large axons (as in the squid), and in muscle cells. But with recent investigation of smaller neurons and processes, membrane potentials ranging down to -20 mV or so have been reported. Although injury is an element in many of these recordings, it is nonetheless becoming clear that a low membrane potential can reflect an important variable in the physiological repertoire of the neuron. The retina provides particularly clear examples of this new concept, and we will discuss later in this chapter and in Chapter 7 how this is related to mechanisms of synaptic transmission.

ACTION POTENTIAL

In our discussion of the parts of the brain (Fig. 1), it was pointed out that one of the most obvious functions of neurons is that of transfer of signals over long distances, from peripheral receptors to the brain, from the brain to muscles and glands, and between

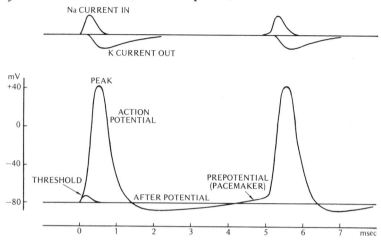

FIG. 8. Sequence of ionic current flows (*above*) and changes in membrane potential (*below*) associated with the generation of action potentials; potential scale in millivolts (mV) at left, time scale in milliseconds (msec) below.

different parts of the brain. This long-distance transfer takes place over long axons by means of *action potentials* or *impulses*. We will describe the mechanism of the action potential as it has been revealed in studies of the largest long-distance axons and then discuss short axons and dendrites.

The characteristics of the action potential are too well known to require detailed comment here (see Katz, 1966; Hodgkin, 1968). The essentials are illustrated in Fig. 8. A small stimulus, in the form of a depolarization of the membrane, causes the membrane to respond with a rapid depolarization, from the resting value of −70 mV to +40 mV or so, following which it rapidly returns to the resting value. Below *threshold*, a stimulus elicits only a local response or no response; above threshold, the membrane goes through its stereotyped depolarization response independent of stimulus intensity. Hence, the membrane response is all or nothing. It is *active*, in that the response reflects a property of the membrane rather than simply the stimulus itself. It is, clearly, a highly nonlinear property.

The essence of this mechanism is that it is regenerative; because it is regenerative it resembles an explosive gas mixture, certain electronic devices, and numerous other physical and chemical systems. The generally accepted model for this mechanism in nerve is that of Hodgkin and Huxley (1952) who, as every student can tell, worked it out in the giant axon of the squid. In simplest terms, the model consists of a positive feedback relationship between Na conductance and membrane depolarization, such that the initial threshold depolarization leads to increased Na conductance and Na ion influx, which causes further depolarization, and so on. This cycle, once initiated, proceeds until the membrane reaches the potential at which there is no more driving force for the Na ions; this is the *equilibrium potential* for Na ions (E_{Na}), about 40 mV inside positive. As this point is reached, Na conductance is turned off (sodium inactivation), and a slower increase in K conductance begins. The E_K is at a slightly more polarized level than the resting membrane potential; hence, the K ion current (together with sodium inactivation) helps return the membrane potential to resting level. The directions of these current flows are indicated in Fig. 7.

The regenerative all-or-nothing property can now be seen as the necessary mechanism for long-distance signal transmission. A patch of membrane undergoing this potential change generates current that acts to depolarize its neighboring patch (see Fig. 7). This patch then goes through exactly the same cycle, and so on, down the length of the axon. The action potential at the end of the axon is identical to that at the beginning; there has been no transmission loss. There have been small exchanges of Na and K along the way, which are restored by the ionic pump. When we speak of *active spread, nervous conduction,* or *impulse propagation,* we mean explicitly transmission by this action potential mechanism.

How then is information transferred by these identical signals? The answer, of course, is that, in the single fiber, it can only be carried as a *frequency code*, reflecting the time intervals between successive impulses. The cycle of generation of a sequence of impulses is illustrated in Fig. 8. An action potential is followed by a slower change, an *afterpotential,* which, in turn, merges with a prepotential from which the next action potential arises. Distinct changes in membrane excitability are associated with each period. During the action potential, the membrane is *absolutely refractory* to further stimulation. During the afterpotential, the membrane is busy pumping back ions, and the threshold is raised, i.e., it is *relatively refractory* to another stimulus. During the ensuing *prepotential,* however, the membrane is hyperexcitable.

In cells that spontaneously generate impulses, the prepotential is also referred to as a *pacemaker potential.* It is an intrinsic property of nerve membrane that is still little understood. The timing of the next impulse is a complex outcome of the excitability cycle, the natural pacemaker properties of the membrane, and the background level of depolarization. The relative frequencies of spontaneous and induced impulses are obviously crucial to the transfer of information by frequency codes.

In transmitting impulses over long distances, speed may be of the essence (as in the muscle reflexes discussed in Chapter 5), and it is enhanced by two factors. One factor is that the larger axons are myelinated; thus, conduction proceeds not continuously in these axons, but jumps, rather, as it were, from node to node. This is *saltatory conduction.* When patches of excitable membrane are thus separated, the separation obviously must not exceed the distance over which current flow along the axon will be sufficient to activate the impulse mechanism. Thus arises the concept of *safety factor,* the excess of current beyond that required for threshold stimulation. The safety factor is normally high, but it becomes an important variable under conditions of high frequency transmission (i.e., through partially refractory membrane), fatigue, anesthesia, injury and disease, and propagation at axonal branch points.

The other factor is that the conduction rate increases with the increasing diameter of the axon; current flows more easily from patch to patch, in accordance with the properties of electrotonus to be discussed in the next chapter. For myelinated axons, the conduction rate, in meters per second, is approximately six times the diameter in microns. In unmyelinated axons, the velocity appears to vary with the square root of the diameter. From the largest myelinated axons of 20 μm diameter down to the finest unmyelinated axons of 0.2 μm, the conduction velocities range from about 120 m/sec down to less than 1 m/sec. There is, thus, about a hundred-fold range of impulse conduction velocities in axons. For the amusement of the intuition, it may be noted that these rates translate to a range of some 200 down to 2 miles per hour.

ACTION POTENTIALS IN THIN AXONS   Thus far we have considered the impulse mechanism in terms of the classical action potential of the giant axon (diameter 0.5-1 mm) of the squid. But, as Alice might have said, there are no squid axons in the brain. Studies of peripheral nerves show that the general model is applicable to the active membrane at the nodes of myelinated axons. We are emboldened, therefore, to assume that the model also applies to axons in the central nervous system. But that is an assumption that must be tested. This is particularly true for the thin unmyelinated axons that make up the majority of axons in the brain and that are the most relevant to synaptic organization.

These axons are the least accessible to experiment, but recent work has begun to reveal their properties. There is evidence, for example, that the density of Na conductance channels in the membrane is surprisingly low. A figure of only 1-2 channels/μm^2 has been estimated for fish olfactory nerves (Colquhoun, Henderson, and Ritchie, 1972); by comparison, the density in squid axons is of the order of 100/μm^2 (Chandler and Meves, 1965). Since the fish olfactory axons have diameters of 0.2 μm, it appears that there is only one channel for every micron or so of length. This has conjured up the vision, to J. M. Ritchie and his co-workers, of a kind of "microsaltatory" conduction taking place from channel to channel along these very thin fibers.

Compared to large axons, a thin fiber has a radically increased ratio of surface membrane to internal volume, and this places a far greater and more immediate burden on the energy-requiring ionic pumps we mentioned above. It has been shown that "following just a few impulses in unmyelinated axons, there are well-defined changes in: the high-energy phosphate compounds required for the recovery process; the electrical activity that reflects the operation of the sodium pump; oxygen consumption; and heat production" (Ritchie, 1971). Unmyelinated axons, consequently, are very dependent on an adequate glucose supply, for use as fuel or possibly as a precursor of energy-providing acetylated compounds.

The nature of the ionic pumping mechanism has been intensively studied. It is a characteristic of unmyelinated fibers that the action potential is followed by a large hyperpolarizing afterpotential, which has been correlated with the rise in internal Na ion. It was first thought that, during this afterpotential, the ionic pump simply ejects one Na ion for every K ion taken in and is, therefore, electrically neutral. There is growing evidence, however, that, to a more or less degree, the Na ion is actively extruded. This results in a net movement of electrical charge across the membrane and, hence, an active contribution to the membrane potential. Such a pump is termed *electrogenic* and is thought to be present in several kinds of nerve cell and nerve process, vertebrate and invertebrate, as well as in a variety of body cells (e.g., muscle cells, red blood cells) (see Ritchie, 1971). This pumping mechanism is important not only for impulse conduction and recovery, but also, in some cases, for synaptic transmission, as will be discussed below.

What of the potassium that appears just outside the membrane during the action potential? It has been found that during the ensuing hyperpolarizing afterpotential the sensitivity of the membrane to K ion is greatly increased over normal. We have therefore this additional factor to consider as a means of interaction between nerve cells and between a nerve cell and its surrounding glial cells.

Thin axons may be long axons, arising from principal neurons and carrying impulses to distant regions; these are the kind of axons that have been the subjects of the studies mentioned above. But thin axons also arise from intrinsic short-axon cells and distribute within their local region. In such cases, the axon may be very short; we will encounter examples (e.g., the thalamus) of axons only a few hundred microns in length, shorter than the dendritic branches of many neurons. Now, if we consider the likely assumption of an impulse duration of 1 msec and a conduction rate of 1 mm/msec, it is immediately evident that the wavelength of the impulse may be greater than the length of the axon. This implies that, at its peak, the impulse would be spread almost equally through most of the length of the axon. This is quite different from the case of the thin axon that projects outside a region, in which case the wavelength of the impulse is only a fraction of the length of the axon. Because the two cases appear so different, it is possible that the impulse in a very short axon may have a significance beyond simple propagation. This will serve to highlight the point that we, in fact, have very little direct information about the physiological properties of the axons of short-axon cells.

ACTION POTENTIALS IN DENDRITES   The properties of dendrites in central neuropil have received a good deal more attention than the properties of short axons. But as long as methods were not available for direct study of activity in central neurons, these properties were only the subject of speculation. An ancient line of thought held that dendrites provide for vegetative functions only. When the action potential in axons came under study, it became fashionable to consider that impulses were the sole means of signal transfer in all parts of central neurons, including their dendrites (see Forbes, 1922). The pendulum swung the other way when the early intracellular studies of Eccles and his collaborators (Eccles, 1953) showed that motoneuron dendrites do not normally generate impulses; these studies and those of Bishop (1957) and others seemed to give credence to the idea that dendrites supported mainly graded activity.

It is only recently, with the introduction of microelectrode recordings combined with intracellular staining techniques and biophysical models, that direct evidence for dendritic properties could be obtained. With these techniques, the question of active versus passive properties of dendrites has become one of the central concerns of neurophysiologists. In reviewing the evidence in later chapters, we will see that spread of signals through some dendrites is by passive means alone, whereas in others there is active spread in addition. It is becoming apparent that it is part of the functional arsenal of dendrites that they may vary in this regard. It is important to realize that in either case there is passive spread and that analysis of dendritic electrotonus, as outlined in Chapter 4, is necessary for understanding active properties.

In closing this section, it may be noted that there is a widely held assumption, deriving from the classical literature, that any process with the structure of an axon must generate action potentials. From this it has been only a short step to defining an axon as an impulse-generating process and any impulse-generating process as an axon. These definitions, however, do not have general validity, a point that has already been touched on in discussing synaptic arrangements in the previous chapter. We will encounter several examples of morphological dendrites that have impulse-generating properties (Chapters 5, 8, and 11), and we will encounter other cells with processes, classically defined as axonal, that do not generate action potentials (Chapter 7). In the previous chapter, we saw that, in many cases, the parts of the neuron do not admit of a simple distinction between axon and dendrite; it should, therefore, not be surprising that the relation between the structural parts of the neuron and their functional properties similarly do not admit of simple generalization. If we define structural parts of the neuron on the basis of structural criteria, and functional properties on the basis of physiological criteria, we will be taking a sound approach to the diversity that characterizes structure-function relations in the brain.

SYNAPTIC POTENTIALS

The action potential is a mechanism for *intra*neuronal transmission, that is, for getting about from one place to another, especially over long distances, within the same neuron. We turn now to the functional properties underlying *inter*neuronal transmission. This takes place over the shortest possible distance, between two apposed membranes. The structure that provides for this transmission is the synapse.

ELECTRICAL SYNAPSES An active patch of membrane generates current that flows in a loop both outside and inside the cell (see Fig. 7). This current can, theoretically, provide for transmission of signals between neurons. It is severely attenuated, however, in leaking out through the extracellular space and crossing the resistance of the neighboring membrane. Signal transmission by this means can, therefore, only occur in special cases, in which large areas of membrane surface are brought together and in which synchronous activity is present. These cases do, in fact, occur, as we shall see: for example, in the Purkinje cell of the cerebellum (Chapter 8).

More effective electrical coupling between cells is achieved through low-resistance connections. These are provided by the *gap junctions* described in the previous chapter. The intercellular channels at these junctions have a very low resistance to current passing between the two neurons, while, at the same time, they prevent loss by leakage to the extracellular space. From this direct connection derive the salient features of electrical synapses. They operate quickly, with little or no delay. They can provide for current flow in both directions, although they can also offer more resistance in one direction than the other (rectification). They provide a means of synchronization of populations of neurons. Their actions can be more fixed and stereotyped, despite repeated use, and less susceptible to metabolic and other effects than chemical synapses (see Bennett, 1972; Nicholls and Purves, 1972).

Electrical synapses are relatively common in invertebrates and lower vertebrates. They are found, particularly, in detection and escape systems, for which their properties (mentioned above) seem to be particularly well suited. They are, as we noted previously, less common in higher vertebrates and mammals. More

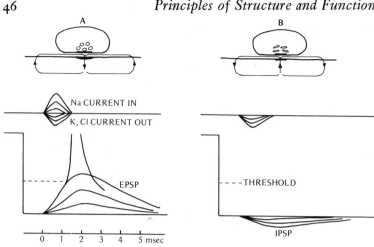

FIG. 9. Basic types of action at chemical synapses. *Above*, pre- and postsynaptic terminals, with net positive current flows shown by arrows for depolarizing (A) and polarizing or hyperpolarizing (B) actions. *Middle*, time course of ionic current flows; note that they are simultaneous rather then sequential, as in the case of the action potential. *Below*, recordings of postsynaptic potentials typical for an EPSP (A) and IPSP (B).

work is needed to determine their incidence and assess their functional role, compared to chemical synapses (see Baker and Llinás, 1971; Bennett, 1972).

CHEMICAL SYNAPSES   The predominant type of synapse in the mammalian brain is the chemical synapse, operating through the release of a transmitter substance from the presynaptic to the postsynaptic terminal. The response of the postsynaptic terminal is called the *synaptic potential.*

There are two basic types of synaptic potential, depending on which way the ions move and whether the membrane, consequently, depolarizes or hyperpolarizes. Our concepts of the mechanisms for the two types are illustrated in Fig. 9 (see Eccles, 1964). To the left, the postsynaptic response to a transmitter substance consists of a net inward movement of positive charge.

Such a movement is brought about by an increase in permeability to Na. Referring again to the membrane patch of Fig. 7, we see that this current flow decreases the polarization across the membrane capacitance (C_m), and, hence, depolarizes the membrane. The equilibrium potential in such a case is characteristically near zero, indicating an increase in conductance to Na, and also to K and other ions. The crucial point is that these ion flows are *simultaneous* and not regenerative; hence, the amount of postsynaptic depolarization is a simple linear function of the amount of transmitter substance liberated, which, in turn, is determined by the amount of presynaptic depolarization. This is indicated in Fig. 9. We say, therefore, that the synaptic is a *graded response*, in contrast to the all-or-nothing response underlying the action potential. In synaptic membranes, the regenerative loop has been opened; the synaptic membrane is, therefore, not "active" in the same sense that the impulse membrane is, and the two must not be confused.

Synapses of this type depolarize the membrane and, thus, are the necessary type for exciting the impulse membrane. These synaptic potentials are, therefore, called *excitatory postsynaptic potentials* (EPSP's), in accordance with the terminology introduced by Eccles and his associates in their pioneering microelectrode studies of the motoneuron.

The second basic type of chemical synaptic response is diagrammed on the right in Fig. 9. The action of the transmitter substance at this junction is to open conductance channels for a net movement of positive charge outward. The equilibrium potential for this ion flow is at a relatively polarized level, say, —80 mV or so, due to increased conductance not only for the outward movement of positive charge (K) but also the inward movement of negative charge (chloride ion, Cl). The path for Cl is also shown in Fig. 8. These ion flows, therefore, tend to hold the membrane potential at a more polarized, or hyperpolarized, level. As in the previous case, this synaptic potential is a graded response. Since synapses of this type tend to keep the membrane from depolarizing, and, hence, work against the initiation of impulses, they are called *inhibitory postsynaptic potentials* (IPSP's). It may be noted that "excitatory" and "inhibitory" have classically been defined relative to impulse initiation. But we will see that some neurons do not have impulse-generating properties (see Chapters 6 and 7). In those neurons, the synaptic response is not converted into impulses; rather, the response leads directly, through electrotonic potential spread, to activation or suppression of synaptic output. In those neurons, therefore, excitation and inhibition are defined relative to synaptic output rather than impulse output. These questions will be considered further in Chapters 6 and 7. For the present, it is important to realize that a more general description of synaptic potentials is in terms of their depolarizing or polarizing (hyperpolarizing) actions, and that the question of whether those actions are excitatory or inhibitory depends on the particular output those actions control.

SYNAPTIC INTEGRATION   It is largely through the interaction between excitatory and inhibitory synapses that the competition for control of the membrane potential in different parts of the neuron is carried out. This competition lies at the heart of the study of the dynamics of synaptic organization. This principle goes back to Sherrington; following him, the process whereby different synaptic imputs are combined within the neuron is termed *synaptic integration*.

The interaction of a single EPSP and IPSP serves as a paradigm for synaptic integration in central neurons, and it will be useful at this point to grasp certain essentials. Let us assume an excitatory synapse and a nearby inhibitory synapse, the activation of which produce the EPSP and IPSP, respectively, as shown in Fig. 10(A). Assume now that the two are activated simultaneously. The effect of the IPSP is to reduce the amplitude of the EPSP, away from the threshold for impulse initiation, as is shown in Fig. 10(A). The dotted line traces the resulting transient; it represents the "integrated" result of the two synaptic potentials.

Now, it is commonly thought that this process of integration is a matter of simple algebraic addition of the two opposed synaptic

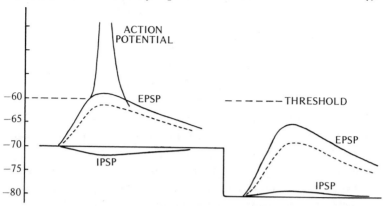

FIG. 10. Illustrations of synaptic integration. (A) The EPSP alone reaches threshold for impulse generation; the IPSP alone causes hyperpolarization; together (dotted line), the IPSP reduces the EPSP below threshold for impulse generation. (B) When the resting membrane is hyperpolarized, the IPSP alone causes depolarization; together with the EPSP, it reduces the EPSP.

potentials; to wit, "depolarization plus hyperpolarization equals membrane potential". This simple formula does not have general validity, however, as is illustrated in Fig. 10(B). When the resting membrane potential happens to be more polarized than the inhibitory equilibrium potential, the IPSP may, in fact, depolarize the membrane (toward the inhibitory equilibrium potential), yet it still brings about a reduction of a simultaneous EPSP. This is because the essential inhibitory action is not a hyperpolarization of the membrane, but rather an increase in conductance for specific ions (e.g., K and Cl) that drives the membrane potential toward the equilibrium potential for those ions, as has been explained above.

It may, thus, be seen that it is the opposition of synaptically induced conductances and ionic currents that controls the membrane potential. The relative amounts of depolarization and hyperpolarization during synaptic integration are a complex outcome of these factors. These factors are in addition to the geometrical relations between excitatory and inhibitory synaptic sites within a dendritic tree, and the electrotonic flow of current between them, to be described in the next chapter.

TRANSMITTER MECHANISMS   Great interest naturally attaches to the identity of the transmitter substances and their mechanism of action at different chemical synapses. This has been worked out most clearly for the neuromuscular junction, thanks to the brilliant investigations of Katz and his co-workers (see Katz, 1966). The junction, or end-plate, is formed by the axon terminals of motoneurons onto skeletal muscles. This type of chemical synapse differs from its counterparts in the brain in several morphological respects: the presynaptic axon terminals are extremely large, covering an area of up to 6000 μm^2; the synaptic cleft is relatively wide (500–1000 Å) and contains a densely staining basal lamina; the postsynaptic membrane (of the muscle cell) forms a trough that receives the axon terminals, the walls of the trough being thrown into numerous folds. This junctional complex is clearly a synapse, but it just as clearly falls at an extreme end of the morphological spectrum. In line with the comments in the preceding chapter, it should be regarded as a giant terminal with a specialized synapse.

A microelectrode suitably placed inside the muscle in the vicinity of an end-plate records, at rest, a membrane potential upon which appear occasional small bumps. Each bump is a *miniature end-plate potential* (miniature EPP), or, colloquially, a "mini". Each mini is the response of the postsynaptic membrane to the release, from the presynaptic terminal, of a discrete packet, or *quantum*, of transmitter substance. There are several lines of evidence that indicate that a quantum is the amount of substance contained in one synaptic vesicle. The rate of release of the quanta is controlled by the membrane potential of the presynaptic terminal; as the membrane is increasingly depolarized, the quanta are released at a faster rate. As we shall see, this is an important factor governing the action of dendrodendritic synapses in the brain.

The synaptic machinery at the neuromuscular junction is acti-

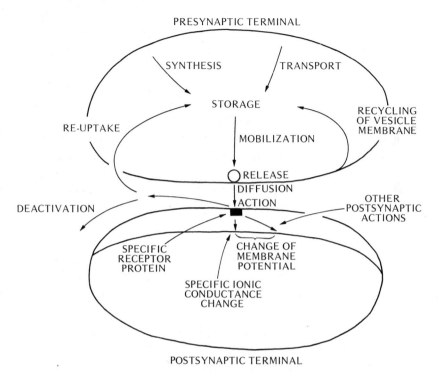

FIG. 11. Diagram of some of the steps that have been identified in the mechanisms of action of chemical synapses.

vated by the all-or-nothing impulse that invades the axon terminals and releases, simultaneously, a large number of quanta, of the order of 100 to 200. There is a synaptic delay of 0.5 msec or so, representing the time it takes the released molecules to diffuse through the cleft and act on the postsynaptic membrane. The postsynaptic response is recorded as a large *end-plate potential*, which is analogous to the synaptic potential of neurons. The end-plate potential triggers the action potential of the muscle fiber, which, in turn, activates the contractile mechanism of the muscle.

As is well known, the work of Katz and co-workers has firmly established acetylcholine as the transmitter substance at the neuromuscular junction. From this work has emerged a picture of the sequence of events, which may be summarized relative to the diagram of Fig. 11. Acetylcholine is synthesized either locally or in the cell body, whence it is transported to the terminal. Within the terminal, it is packaged in vesicles to form a storage moiety. Then, the vesicles move to the synaptic site; it is here that they are regarded strictly speaking, as "synaptic vesicles". Depolarization mobilizes the vesicles and causes them to open onto the synaptic cleft and there to deposit their contents of acetylcholine. The acetylcholine diffuses the short distance to the postsynaptic membrane and acts on specific acetylcholine receptor molecules in that membrane. The effect is to increase the conductance and depolarize the membrane, as already noted. At a slower rate, the acetylcholine is enzymatically split, and much of the choline is taken up again for the synthesis of new transmitter. Current work indicates that the used-up vesicle fuses with the membrane of the presynaptic terminal and that new vesicles are formed from the membrane at the margins of the terminal (Heuser and Reese, 1973; see also Douglas, Hagasawa, and Schulz, 1971). Here, indeed, is an efficient factory, which bottles and markets the transmitter and recycles the juice and the glass.

How does this mechanism at the neuromuscular junction compare with that at other peripheral synapses? In the autonomic nervous system, a basic distinction is made between synapses that operate with acetylcholine (cholinergic synapses) and those that operate with catecholamines (adrenergic synapses). The sequence of steps in the action of adrenergic synapses is analogous to that illustrated in Fig. 11. There are a number of differences related to the metabolic pathways for the two substances, and there is the particular difference that noradrenaline is not mainly taken up again into the presynaptic terminal. These points are reviewed in standard textbooks.

Recently, there has been much work on the possible role of cyclic $3',5'$-adenosine monophosphate (cyclic AMP) as a mediator of adrenergic effects. The general scheme, at present, is that the transmitter substance activates the enzyme adenyl cyclase in the postsynaptic membrane that synthesizes cyclic AMP and causes the postsynaptic response (cf. Greengaard, Kebabian, and Mc-Afee, 1972). These steps are summarized under the heading "other postsynaptic actions" in the diagram of Fig. 11.

In the heart, the inhibitory action of the vagus nerve is mediated by acetylcholine. We have just described synaptic *excitation* of skeletal muscle by acetylcholine, so we immediately recognize an important principle of synaptic organization, that *a specific transmitter substance cannot be identified with a specific postsynaptic action.* In more general terms, the nervous system can use the same substance for different purposes; obversely, the same function (i.e., excitation) can be mediated by different substances. This flexible relation between transmitter substance and physiological action may be regarded as a biochemical corollary to the flexible relation between structure and function of neuronal processes in different parts of the brain.

DALE'S PRINCIPLE   Although the nervous system as a whole can use different substances at different synapses, this is not necessarily true for an individual neuron. The metabolic unity of the neuron would seem to require that it release the same transmitter substance at all its synapses. This is *Dale's Principle (Law)* and, since it can be easily misunderstood, it is well to quote the original formulation. In a review of synaptic transmission in the autonomic nervous system many years ago, Dale, one of the great pioneers in this field, wrote

. . . the phenomena of regeneration appear to indicate that the nature of the chemical function, whether cholinergic or adrenergic, is characteristic for each particular neurone, and unchangeable. When we are dealing with two different endings of the same sensory neurone, the one peripheral and concerned with vasodilatation and the other at a central synapse, can we suppose that the discovery and identification of a chemical transmitter of axon-reflex dilatation would furnish a hint as to the nature of the transmission process at a central synapse? The possibility has at least some value as a stimulus to further experiment. (Dale, 1935.)

This is the acorn from which the mighty oak has grown. The principle is profound, for it implies, as Iversen (1970) has pointed out, that during development some process of differentiation determines the particular secretory product a given neuron will manufacture, store, and release. The usefulness of the principle in the analysis of synaptic organization is explicit in Dale's statement, for, if a substance can be established as the transmitter at one synapse, it can be inferred to be the transmitter at all other synapses made by that neuron.

The point that is often misunderstood is that Dale's Law only applies to the presynaptic unity of the neuron; it does not apply to the postsynaptic actions the transmitter will have at the synapses made by the neuron onto different target neurons. These actions may be similar, or they may be different. We have already noted that the acetylcholine released at motoneuron nerve terminals has an excitatory action at the neuromuscular junction, whereas acetylcholine released from vagal nerve terminals has an inhibitory action in the heart. Similar possibilities for diversity of action exist for the same transmitter released from the different terminals of the same neuron. This has been brought out clearly by recent experiments in invertebrate nervous systems. In molluscs (e.g., the snail, *Aplysia*), a single neuron has been found that produces EPSP's in one type of postsynaptic neuron and IPSP's in another; acetylcholine is the transmitter at both synapses (Tauc and Gerschenfeld, 1961). Thus, an important corollary to Dale's Law is that the action of a transmitter substance is determined by specific mechanisms at the particular postsynaptic sites. We shall return to this theme shortly.

TRANSMITTER MECHANISMS IN CENTRAL SYNAPSES   Workers in several disciplines—biochemistry, neuropharmacology, electrophysiology, histochemistry—have, over the past generation, followed the transmitter trails into the central nervous system. Their work has supplied evidence that, in addition to cholinergic and adrenergic synapses, there are synapses operating through amino acid mediators. Glutamic and aspartic acid have been implicated at excitatory synapses, whereas gamma-aminobutyric acid (GABA) and glycine have been implicated at inhibitory synapses. If we keep in mind the flexible relation that may exist between a chemical substance and its physiological action, and realize the difficulties of working with central nervous tissue, we can see that the task of sifting the evidence for these and other putative transmitters is a very difficult one indeed. A useful guide is the recent review by Cooper, Bloom, and Roth (1974).

In pursuing transmitter mechanisms in central synapses, it seems safe to predict that variations will be found in the classical picture drawn from peripheral junctions. Referring again to the equivalent circuit of Fig. 7, we can see that a depolarization of the membrane can be brought about, not only by an increase in Na conductance (as at the neuromuscular junction), but also by a decrease in K conductance. Such a mechanism has, in fact, been postulated for the generation of an EPSP in frog sympathetic ganglia (Weight and Votova, 1970). Conversely, a hyperpolarization can be brought about, not only by an increase in K and/or Cl conductance, but also by a decrease in Na conductance. The latter mechanism has also been postulated in the sympathetic ganglion (Weight and Padjen, 1973). A similar mechanism appears to be involved in synaptic transmission in the retina, as will be discussed in Chapter 7.

Recently, it has been suggested that some synapses may act by a mechanism other than a conductance increase in the postsynaptic membrane (Krnjevic, 1970). It is envisaged, instead, that an amino acid (glutamate, for example) released by the presynaptic terminal attaches to a carrier (protein molecule?) in the postsynaptic membrane. The attachment greatly increases the affinity of the carrier for Na, and the entire complex is then driven across the membrane by the Na gradient. The net transfer of cation depolarizes the membrane, resulting in an EPSP. For such a synapse, the upper curve in Fig. 9 would still describe the time course of Na movement, but by carrier rather than conductance change, whereas the EPSP response would be described as before by the lower curve. Whether or not this can be proved to be the mechanism at a given central synapse, it is of general interest because there is evidence that similar mechanisms operate elsewhere in the body, as in the uptake of amino acids from the digestive tract (cf. Schultz and Curran, 1970).

It can be seen that, although the actions of synapses may be simply characterized as depolarizing or hyperpolarizing, the mechanisms by which they achieve those actions are quite various. Our main concern will be limited to the synaptic actions, and our conclusions about their significance will, in general, be independent of the details of the mechanisms. The analysis of synaptic mechanisms, quite apart from the problem of identifying the transmitter substances involved, is a difficult subject, for which sophisticated electrophysiological methods are required; for an introduction to these methods, the reader is referred to the monographs by Eccles (1964) and Hubbard, Llinás, and Quastel (1970).

Because of the close link between metabolism and transmitter substances, the intensity and duration of synaptic activity is an important variable in determining synaptic efficacy. Studies of the neuromuscular junction have delimited periods of facilitation and suppression that follow an initial period of high frequency stimulation (cf. Martin, 1967). Facilitation and depression have also been demonstrated at a number of central synapses. The nervous systems of invertebrates are useful models for the study of these properties. In neurons of the leech, for example, a hyperpolarization occurs following impulse activity that appears to be due to an electrogenic sodium pump like that in fine axons (Baylor and Nicholls, 1971; see also previous section). It has been shown that this hyperpolarization has profound effects on synaptic transmission in the neuropil: increase in EPSP amplitude, decrease in IPSP's, conduction block in dendritic processes, and, probably, changes in amount of presynaptic transmitter release. In the snail, there is evidence for a conditioning of synaptic transmission and changes in reflex behavior (Kandel and Spencer, 1968). There is also tantalizing evidence that synaptic activity affects RNA synthesis in the snail (Kernell and Petersen, 1970; Berry and Cohen, 1972).

It is evident from these considerations, and from the diagram of Fig. 10, that the mechanism of synaptic transmission is not such a simple, stereotyped, self-contained type of affair as that of an action potential. There are obviously many steps involved in a synaptic action, and the steps are of such differing natures as metabolic, biophysical, electrical. For this reason, the synapse is the most vulnerable link in the neuronal circuits, the most sensitive site for modifying transmission.

It is obvious that if modifications dependent on use or disuse occur at the central synapses of the vertebrate brain, they would provide a basis for changes underlying learning and memory. But because these changes can often be demonstrated only under extreme or artificial conditions, at peripheral and invertebrate junctions with no natural concomitant of learning or memory, caution is necessary in interpreting them.

Within the brain itself, synapses are, of course, modifiable during the differentiation and growth of neurons in embryonic and early life; the processes concerned, however, remain among the most profound unsolved problems of biology. In the adult brain, there is more and more experimental evidence of the modifiability of synapses. Rearrangements of synaptic inputs have been found following lesions of certain pathways (Raisman, 1970), and we will discuss the effect of de-afferentation on dendritic spines in the cerebral cortex in Chapter 12. There is also evidence that quite dramatic changes in synaptic potency occur during and after repetitive activation of the hippocampus (Chapter 11) and neocortex (Chapter 12), similar to some of the effects elicited in neurons of the leech, as noted above. These effects, like those in invertebrate preparations, have been produced artificially, and the evidence for such changes as a basis for behavior must, thus, remain tentative. Nonetheless, it is widely accepted that synaptic changes do underlie learning and memory. The study of synaptic organization is, therefore, necessary for an understanding of these aspects of vertebrate behavior.

4

DENDRITIC
ELECTROTONUS

Dendrites provide an elaboration of the surface area of a neuron, and synapses characteristically have different sites on this surface. The study of synaptic organization is, therefore, a study not only of action at a given synapse, but also of spatial relations between that synapse and the dendritic tree. These relations govern the effectiveness of a synaptic potential in generating a further output, whether that output be through a nearby dendrodendritic synapse or through an impulse-generating membrane within the dendritic tree or a soma or an axon hillock. These functional interactions are mediated by the flow of electrical currents set up by the synaptic potentials, and the task is to describe these current flows as rigorously as possible.

The analysis of current flows in neuronal processes has a long tradition in neurophysiology and biophysics. It began with the very earliest electrophysiological studies of activity in peripheral nerves in the 1840's. Much subsequent work in the 19th century was devoted to the problem of distinguishing the passive spread of current in peripheral nerves from the "action" currents associated with the action potential. From this work arose the concept of the nerve fiber as a core conductor, deriving from its high membrane resistance and relatively low internal resistance. Passive spread through the core conductor was termed *electrotonus*, and the changes in membrane potential caused by the passive spread were termed *electrotonic potentials*. These current flows and potential changes were described mathematically, and it was soon realized that the equations were essentially similar to those for the flow of current in an electrical cable, the flow of heat in a rod, and the diffusion of substances in a solute. The equations as applied to nerve cells are often referred to as the *cable equations*.

Application of the cable equations to experiments on the single nerve axon awaited the modern era, beginning with the analysis of the action potential mechanism and the development of the Hodgkin-Huxley model. The equations were adapted for use in studying conduction of the action potential in muscle fibers, and they were also important in Katz's analysis of the end-plate potential at the neuromuscular junction.

The application of the equations to these relatively simple cases (in a geometrical sense) was no easy task, and it, therefore, seemed that neuronal dendrites, with their limitless variety of shapes and branching patterns and their inaccessability to experimental approaches, were beyond the reach of rigorous analysis. Beginning in 1957, however, the systematic adaptation of the equations to dendrites has been carried out by Rall. Some of the key papers in this series are listed in the bibliography for the benefit of the biophysically minded student (Rall, 1957, 1959a,b, 1962a,b, 1964, 1967, 1969, 1970a,b). We will describe here only the essentials of the methods, in order to provide the student with the basic tools for understanding the spread of synaptic responses and the functional properties of dendrites. For further orientation to the mathematical methods, the reader is referred to the recent monograph of Noble, Jack, and Tsien, *Electric Current Flow in Excitable Cells* (1974).

STEADY-STATE ELECTROTONUS

The basic characteristics of electrotonic potentials are illustrated in Fig. 12. We consider a length of neuronal process (either axon or dendrite) as a series of membrane patches. A steady depolarization is imposed on the membrane of patch 1. This places a per-

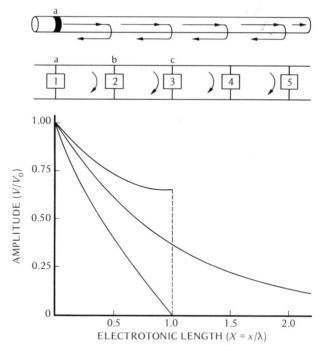

FIG. 12. Electrotonic currents and potentials. *Above*, a nerve process (axon or dendrite) with a steady depolarization imposed at (a); arrows, resultant current flows. *Middle*, current flows for a membrane patch model of nerve process. *Below*, distribution of electrotonic potential for the case of infinite length (continuous line); for termination of a process in an open (short-circuit) ending at $X = 1$ (lower line); for termination of a process in a closed ending at $X = 1$ (upper line). *Ordinate*, change in membrane potential (V) relative to its value at the site of current injection (V_0). (After Rall, 1959a.)

sisting excess of positive charge inside the membrane of patch 1. In this *steady-state* condition, the membrane capacitance acts as an open circuit, and we need only consider, therefore, the current flow through the resistance pathways of the neuron. Some of the current at point (a) will leak back through the membrane resistance of patch 1, but some will flow through the small internal resistance to point (b). At point (b), the current has two similar alternative paths: back through the membrane resistance of patch 2 or onward through the small internal resistance to point (c). The sequence repeats itself further along the neuronal process until all the current has leaked back across the membrane and returned to patch 1 to complete the circuit.

From the above, it can be seen that there will be a gradient of current flow along the fiber and a corresponding gradient of membrane potential. The difference between the membrane potential (V_m) and its resting value (E_r) at any point along the fiber is the *electrotonic potential* (V); $V = V_m - E_r$. For the idealized case of Fig. 12, we require that the electrical properties of a given patch do not change (as they do during action potentials and synaptic potentials). The properties are constant or, as we say, passive. Electrotonic potentials are, therefore, also called *passive potentials*.

If the electrical properties of the neuronal process are known, then, as the student familiar with Kirkhoff's elementary laws of electrical circuits knows, it is possible to describe the flow of current and the distribution of electrotonic potential exactly. The equation, as it applies to a neuronal process, is one of the so-called *cable equations*

$$V = \frac{d^2 V}{d X^2} \tag{1}$$

in which V = electrotonic potential, and X = length in electrotonic units, which we will define shortly.

One solution of this equation gives

$$V = V_0 \, e^{-x/\lambda} \tag{2}$$

in which $V_0 = V$ at $x = 0$ (the starting point on the neuronal process), and λ = characteristic length. The significance of λ can be seen by setting $x = \lambda$; then $V = 1/e$ or, approximately, 0.37 of V_0. Thus, λ is the length over which the electrotonic potential falls to $1/e$ of its value at the origin. Like the notion of half-life, it is a way of characterizing and comparing electrotonic spread in different kinds of neuronal processes. It is a concept that is central to the analysis of dendritic organization.

A plot of Eq. (2) is shown in Fig. 12 (middle trace of bottom graph). The convention we will employ is to plot V relative to V_0 on the ordinate. On the abscissa is plotted the electrotonic length X, which is real length (x) relative to characteristic length λ, i.e., the exponent of Eq. (2).

To use Eq. (2) to describe the spread of synaptic potentials, we must take into account the geometry of the particular dendrites under study and their electrical properties. The basic geometrical variables whose numerical values we must determine are the *length*, the *diameter*, and the *branching* of the dendritic processes. Let us now consider these three aspects.

DENDRITIC LENGTH   One of the assumptions underlying Eq. (2) is that the cable is of infinite length. For long axons, this assumption is acceptable, for virtually all the current has leaked back over several characteristic lengths (for example, V/V_0 is less than 0.02 at $X = 3$). Dendrites are relatively short, however, and the equation must obviously take this into account. This requires not only knowledge of the length of a dendrite, but also of its mode of termination, a "boundary condition" in mathematical terminology.

An intuitive grasp of the importance of these factors is provided by two cases illustrated by the dotted line gradients in Fig. 12. We assume a dendrite of electrotonic length $X = 1$. In one case, the dendrite is assumed to have an open ending (i.e., $V_m = E_r$ at $X = 1$). At that point, therefore, the electrotonic potential must be 0, and associated with this is a much steeper gradient of electrotonic potential along the dendrite, as is shown by the lower line in Fig. 12. This case is relevant to situations in which a synaptic potential spreads toward a site of high conductance, as, for example, a strong inhibitory synaptic site, or a confluence of many branches. In such situations, the electrotonic gradient is steeper, and the spread of electrotonic potential is less (although usually not to the extent produced by the complete short circuit used for this illustration).

On the other hand, the dendrite may be assumed to terminate in a normal patch of membrane. This case will obviously cause the electrotonic potential to be greater at this point than it would be if the fiber extended to infinity. It can be shown that this is closely approximated by the assumption of a patch of infinite resistance across the end of the fiber, an assumption that is easier to work with mathematically. The electrotonic potentials along the dendrite, for this case, are shown by the upper line in Fig. 12. This case has general applicability to situations in which a synaptic potential spreads through a dendritic branch, as well as through an entire dendritic tree. In these situations, the electrotonic gradient is less, and the spread of electrotonic potential is greater than for an infinite cable.

DENDRITIC DIAMETER   All neuronal processes, including dendrites, vary in diameter as well as length. Intuitively, it is obvious that electrotonic spread will depend on diameter. Spread of current will be enhanced by a larger diameter, since the effective resistance inside a neuronal process (r_i) decreases as the diameter of the process increases. Similarly, spread will be enhanced by a relatively high membrane resistance (r_m), so that the current tends to spread internally through the process rather than across the membrane. These same factors indeed underlie the greater velocity of the action potential in larger fibers.

The dependence of electrotonic spread on diameter is incorporated in the equation for the *characteristic length* (λ). When the resistances are expressed in terms of unit area, the equation is

$$\lambda = \sqrt{\frac{r_m}{r_i}} = \sqrt{\frac{R_m}{R_i} \cdot \frac{d}{4}} \qquad (3)$$

where $r_m = R_m/2\pi a$ = membrane resistance for unit length of cylinder ($\Omega\,$cm), $r_i = R_i/\pi a^2$ = internal resistance for unit length of cylinder (Ω/cm), R_m = specific resistance of unit area of membrane ($\Omega\,$cm^2), R_i = specific resistance of unit volume of internal medium ($\Omega\,$cm), a = radius of cylinder (cm), and Ω = resistance in ohms. In other words, the characteristic length (λ) varies as the square root of the diameter, given fixed values for R_m and R_i.

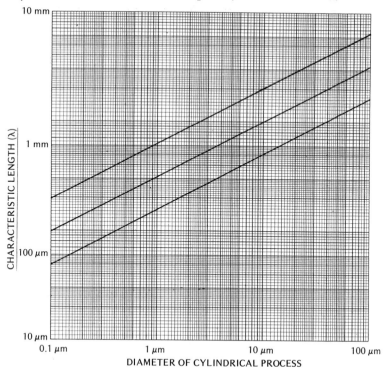

FIG. 13. Dependence of the characteristic length (λ) on the diameter and electrical parameters of a nerve process. *Ordinate*, characteristic length. *Abscissa*, diameter of a nerve process. The lines plot

$$\lambda = \sqrt{\frac{R_m}{R_i} \cdot \frac{d}{4}}$$

for three values of the ratio R_m/R_i: 10, 40, and 160 (e.g., $R_i = 50$ Ωcm; $R_m = 500$, 2000, and 8000 Ωcm^2).

The variation of the characteristic length with the square root of the diameter is another key concept in the study of synaptic organization and the properties of dendrites. This relationship is illustrated in Fig. 13. In the construction of this nomogram, assumptions about the values of the electric properties R_i and R_m were required; R_i is relatively constant, and has been estimated at approximately 50 Ωcm for mammalian nerve, but R_m is more variable, and three values have, therefore, been used to cover most of the range found in mammalian nerve membrane.

Estimation of the specific membrane resistance (R_m) is by no means a simple procedure. One begins by passing current through an intracellular electrode and simultaneously recording the change in potential this produces. The current-voltage relation yields a value for the *whole neuron resistance* (R_N), also called the *input resistance*. The question arises: How much of the resistance (or, inversely, conductance) is due to the current that flows across the soma membrane and how much to that flowing out through the dendrites? This requires an assessment of the branching pattern of the dendritic tree (see below), from which the relative contributions of *dendritic* and *somatic input conductance* are obtained, and expressed as the ratio (ρ). Since the ratio is usually greater than 1, it expresses *dendritic dominance*, in an electrotonic sense, just as measurements of surface area express dendritic dominance, in a spatial sense. An expression that includes ρ and various other factors is then used to obtain an estimate for R_m. These methods are explained by Rall (1959b, 1960), and the procedure, as it applies to the analysis of the motoneuron, is succinctly described by Lux, Schubert, and Kreutzberg (1970).

For general orientation, Fig. 12 provides the student with a starting point in assessing the range of possibilities for spread of currents in dendrites of different diameters. We will have occasion to refer frequently to this graph in our studies of the dendritic properties of different neurons. By way of general comment, it may be noted here that λ has a relatively high value even in very fine processes. Thus, the finest dendritic branches in the brain are of the order of 0.1 μm in diameter; as can be seen in Fig. 13, one might expect a λ of the order of 100 μm for them. Such branches are, in fact, rarely this long, which indicates that current spread may well be rather effective in small dendritic trees. On the other hand, the largest dendritic trunks (as, for example, in motoneurons and cortical pyramidal cells) are 10-15 μm in diameter. From Fig. 13 we might expect a dendrite of

10 μm in diameter to have a λ of the order of 1000 μm (1 mm). This is rather longer than the length of most large dendrites (but see the neocortex, Chapter 12), which again indicates that the spread of synaptic current through them may be relatively effective. In contrast, the large axons of principal neurons are many times longer than their characteristic lengths; the longest motoneuron axons to the lower limb, for example, are over a meter (1000 mm) in length. Communication through these axons must be by action potentials rather than by passive potentials alone.

DENDRITIC BRANCHING The third basic characteristic of dendritic geometry is branching. Dendrites display a fantastic variety of branching patterns. How do the methods of Rall deal with this fact?

The basic approach is illustrated in Fig. 14. We consider, as usual, the simplest case first, that of a dendrite that divides into two branches. A steady depolarization is set up in the terminal patch of each branch, as might be due, for example, to steady excitatory synaptic potentials at those sites. In each branch, there is an electrotonic current flow, as depicted in Fig. 14. At the branch point, the currents summate and spread into the stem dendrite.

Now if both branches have the same diameter, length, and electrical properties, the gradient of electrotonic potential will be the same in them, and the gradient in one branch will be representative for both. We may then combine the branches into a single cylinder that represents the electrical properties, current flow, and potential spread for both branches. Thus, we obtain an *equivalent cylinder* for these branches: the length of the cylinder in Fig. 14 from $x = 0$ to $x = a$.

What will be the gradient of electrotonic potential in the equivalent cylinder? This depends on the diameters of the branches, of course, but it also depends on the relations of the branches to the stem fiber. In other words, there is a boundary condition at the branch point. Let us imagine the simplest case, in which there would be continuity of the gradient through the

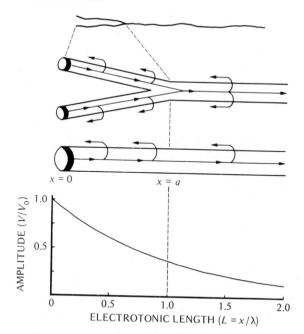

FIG. 14. Derivation of an equivalent cylinder for electrotonic current flow in a nerve process with two branches. *Steps from top to bottom,* branching process, cable model, equivalent cylinder. Steady depolarization imposed at $X = 0$; branch point is at (a); arrows, current flows. *Below,* continuous decrement of electrotonic potential, assuming a 3/2 power constraint at branch point, as defined in the text.

branches and into the stem fiber. This is plotted in Fig. 14. Rall has shown that this occurs if the branch diameters, taken to their 3/2 power and added, equal the 3/2 power of the diameter of the stem fiber.[1] This fulfills the requirement that the combined conductance for a given electrotonic length of the branches equals

[1] The dependence of dendritic input conductance (G_D) on the 3/2 power of the diameter of a dendrite of semi-infinite extension was derived by Rall (1959b) as follows

$$G_D = \frac{I}{V} \quad \text{(Ohm's Law)}$$
$$= (\lambda r_i)^{-1}$$
$$= \left(\frac{\pi}{2}\right)\left(\frac{1}{R_m R_i}\right)^{1/2}(d)^{3/2} \qquad (4)$$

the conductance for the same electrotonic length of the stem fiber. Using this assumption, we can fit the equivalent cylinder for the branches to the cylinder for the trunk, as in Fig. 14, and obtain, thereby, a single cylinder for the entire dendritic system.

As a step in the construction of the equivalent cylinder, we have to generalize *electrotonic length* (X) in order to apply it to the combined branches; thus, $L = x/\lambda$ applies to both single dendrites and to parallel branches, and L will be the designation we will use for electrotonic length hereafter.

Let us take a practical example. Assume that the branches in Fig. 14 are each 1 μm in diameter. This value raised to the 3/2 power is also 1; the sum for the two branches is, therefore, 2. The value for the stem dendrite which, raised to the 3/2 power, equals 2, is approximately 1.6. Thus, if the diameter of the stem dendrite is 1.6 μm, there will be electrotonic continuity at the branch point. If we make some reasonable assumptions about the electrical properties (taking, for example, the middle value for R_m in the graph of Fig. 13), we can then derive the characteristic lengths for the stem fiber and the branches and, finally, estimate the gradient of electrotonic potential through the system, as shown in Fig. 14. Note that, as a reflection of its larger diameter, the stem dendrite has a λ of about 400 μm, in comparison with 300 μm for the branches. This means that there is a continuity of the electrotonic gradient with respect to the electrotonic length (L) but not with respect to the actual length (x); the gradient is less steep in the stem dendrite with respect to the real length. The student should be sure he understands this difference.

Just as the characteristic length provides a means for comparing spread of synaptic potentials in single dendrites of different size, so does the concept of equivalent cylinder provide the means for comparing spread in branching dendritic trees of different size. In subsequent chapters, we will derive equivalent cylinders, where possible, for dendritic branching systems by the steps outlined above.

It may be asked whether the 3/2 power relationship between a dendritic process and its branches has a general validity in the brain. This turns out to be, in fact, a reasonable first approximation for a number of dendritic trees. The equations are simple under this constraint (because the cylinder has a uniform radius throughout its length), but any branching relationship can be incorporated into the equivalent cylinder model with suitable assumptions. To emphasize the general applications, one may refer to it simply as an *electrotonic model* for whatever dendritic system one is interested in.

TRANSIENT ELECTROTONUS

Thus far we have described the spread of electrotonic current under steady-state conditions, that is, assuming a steady synaptic input. This is a necessary starting point for characterizing the electrotonic properties of dendrites. There are, indeed, cases in local brain regions of background depolarizations (as in the visual receptors of the retina) that are essentially steady states. There are also many types of slow synaptic potentials for which the assumption of a steady state is reasonable.

There are, however, many cases in which the essence of synaptic action is rapid transmission between neurons, an immediate triggering of an impulse within the same neuron, or both. The synaptic potentials for these functions have a rapid onset and decay; the description of these rapid transients requires an elaboration of the cable equations that is the biophysical foundation for the study of *transient electrotonus*.

Conceptually, the analysis of transient electrotonus is simple enough; it involves putting the capacitance of the membrane back into the picture. Rapid changes in electric charge across the membrane must first charge or discharge the membrane capacitance. This results in a time-dependent storage of electrical energy that delays, distorts, and attenuates the imposed signal. This is seen clearly in experiments in which a step of current is introduced through an intracellular electrode, and the voltage change is recorded through the same electrode. The voltage rises more slowly; we call it a *charging transient*. The time to reach a value $(1/e)$ of the final level is called a *charging time constant* or, alternatively, the *whole neuron time constant* (τ_N). As in the case of the whole neuron resistance described above, it is necessary to take into ac-

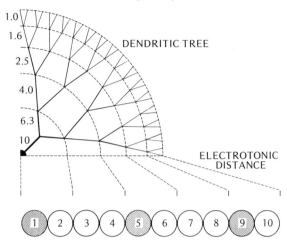

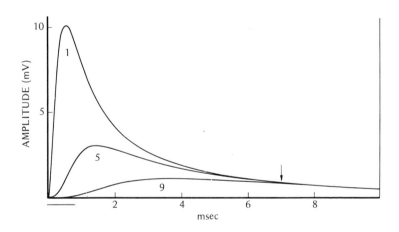

FIG. 15. Transient electrotonus in a dendritic tree. *Above,* diagrammatic representation of an extensively branched dendritic tree, in which a 3/2 power constraint applies to all branch points. *Middle,* equivalent cylinder for a dendritic tree, modeled as a chain of compartments (patches). *Below,* transient electrotonic potentials that would be recorded from the cell body (1) for the cases of brief synaptic conductance change (bar) in compartments (1), (5), and (9). *Ordinate,* amplitude (mV). *Abscissa,* time (msec). (Above and middle diagrams after Rall, 1964; bottom tracings calculated from the Rall model on a PDP-10 computer by K. L. Marton.)

count the electrotonic properties of the dendritic tree before an estimate of the membrane time constant (τ_m) can be made. These methods are described in Rall (1969, 1970b).

The membrane time constant, together with the electrotonic properties of the dendritic tree, determines the time course of a charging transient or a synaptic potential. Mathematically the description of these relationships is quite complicated. The essence of the methods, as they apply to the analysis of synaptic potentials, may be illustrated with respect to the diagram of Fig. 15. Consider a cell with an extensively branched dendritic tree, in which we wish to compare the synaptic potentials recorded in the cell body when the synaptic inputs are at three different sites: the cell body itself, midway in the dendritic tree, and at the far end of the tree. What will be the effect of the electrotonic properties of the dendrites on the characteristics of the synaptic potentials as they reach the cell body?

To make this case tractable, it is assumed that each dendrite gives rise to two daughter branches that meet the 3/2 power constraint previously described. By suitable assumptions for the branch lengths, an equivalent cylinder is obtained in which increments of electrotonic length (ΔL) correspond to successive levels of branching. This modeling of an extensively branched system illustrates the power of the equivalent cylinder concept more vividly than the previous simple case.

For computational purposes the equivalent cylinder is put into the form of a chain of compartments, as shown in Fig. 15(B). Each compartment represents a patch of membrane containing the simple electrical circuit of Fig. 12. Each patch is joined to its neighbor by transfer constants that incorporate the electrotonic properties of the dendrites. It can be seen that this approach is essentially similar to the compartmental analyses that are widely used in the study of metabolic turnover in many systems of the body.

To model a synapse in this system, we assume a brief conductance change, which places an excess positive charge inside the synaptic membrane. This depolarizes the membrane, giving rise to an EPSP, as in our previous simple model (Fig. 9). We assume the three different sites of synaptic input shown in Fig. 15 and ask the computer to compute the resulting voltage transients in the cell body due to electrotonic spread from these sites.

The results are shown in Fig. 15(C). When the synaptic locus is in the soma, the EPSP at that site has an immediate and sharp rise and a rapid decay. For the middle dendritic input, the soma transient is reduced in amplitude and has a slower time course. This effect becomes more extreme with the peripheral input site. In these respects, the results agree with our intuitive expectations. The computer, however, allows precise measurements of these different EPSP's to be made and compared. For example, Rall has shown that certain kinds of *shape index*, such as the time-to-peak and the half-width, are sensitive indicators of the sites of input. This has been a valuable tool in the analysis of synaptic loci in motoneurons (Chapter 5).

The model also permits one to be quite precise in describing synaptic current flows. For the soma input, for example, we can say that the decay of the EPSP is more rapid than would be the case for a single RC element (i.e., membrane patch) because, in addition to leakage of the charge locally across the membrane, there is a rapid electrotonic spread of the charge from that patch to the rest of the cylinder. Rall conceives of this as *equalizing spread*, for which *equalizing time constants* can be obtained through standard techniques of peeling exponentials from semi-logarithmic plots of the decays (Rall, 1969, 1970b). As time goes by the charge gets evenly distributed throughout the cylinder, as is shown by the fact that the final decays of all the transients, regardless of their initial input locus, become similar [beginning at the arrow in Fig. 15(C)].

Consider now the soma EPSP generated by the distant dendritic input. Not only is this transient slow and attenuated, but it has a very slow onset (i.e., a long time-to-peak). In fact, during the time of the synaptic conductance change in the periphery, no change in soma membrane potential is observed at all! This, of course, is a property of dendritic electrotonus, in that it takes time for the charge to spread along the capacitance of the equivalent cylinder.

The general conclusion from this type of analysis is that local inputs provide for rapid responses, whereas distant inputs provide for slower and more sustained activity. The rapid responses may have the function, in some neurons, of triggering impulses (as in the motoneuron of the spinal cord). In other neurons, their function may be the activation of local synaptic output from the same dendritic region (as in olfactory bulb neurons). Any local input will also contribute to the slow modulation of the background excitability of distant sites within its dendritic tree. It is important to realize that, because of the dendrodendritic synapses present in many brain regions, synapses onto small dendritic branches, although distant with respect to the site of impulse initiation in the soma, may be local with respect to postsynaptic interactions and to dendrodendritic synaptic output.

SYNAPTIC INTERACTIONS We previously discussed the interaction between an EPSP and an IPSP at the same site on a dendrite, and we now must consider the effect different sites have on synaptic interactions. An example that has been studied by Rall (1964) is illustrated in Fig. 16. We assume a dendritic cylinder divided into five compartments. The recording site is at 1. An excitatory input (e) occurs in the middle of the dendrite, and the resulting EPSP recorded at 1 is shown by the control tracing in the diagram. Note that the EPSP peak occurs well after the excitatory input occurs. This, of course, reflects the time required for electrotonic spread from the input site to the recording site.

Relative to this excitatory input, we have three basic possibilities for the site of the inhibitory input (i). One site is distal, further out on the dendrite (example A). Another is midway, at the same site as the excitatory input (example B). The third is proximal, at the recording site (example C). What is the effect on the recorded EPSP of a continuous inhibitory input at these three sites?

The computed results are shown in Fig. 16. The distal IPSP

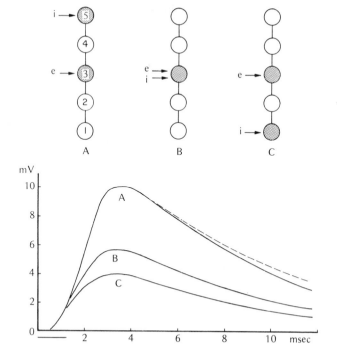

FIG. 16. Dependence of synaptic integration on the electrotonic relation between input sites. In all cases an excitatory synapse (e) is located in the middle of an equivalent cylinder (position 3). An inhibitory synapse (i) is located at one of three possible sites: peripherally (A), same site (B), or proximally (C). *Below,* resulting synaptic potentials recorded proximally (at 1) for these three cases. Dotted line, control response to an excitatory synapse alone. (After Rall, 1964.)

(A) has practically no effect on the peak of the EPSP but does cause it to decline more rapidly. The middle IPSP (B) cuts the EPSP peak to 57% of the control value. The proximal IPSP (C) is even more effective; it cuts the EPSP peak to 40%. We may remind ourselves that these interactions do not reflect an algebraic summation of the excitatory and inhibitory potentials. They are the outcome, rather, of the interactions of current flows induced by the synaptic conductance changes and then modified by the electrotonic relations between the sites.

This simple example will serve as a useful model for many of the cases of synaptic integration that we will encounter in local brain regions. It will apply, for instance, to the case of synaptic interactions within a single branch of a dendrite. For such a case the recording site might be located near a dendrodendritic synaptic output from this dendrite, and the synaptic potentials spreading to that site might determine, in a graded manner, the output from that synapse. On the other hand, the simple chain of compartments depicted in Fig. 16 could be used to represent, with suitable assumptions, the equivalent cylinder of Fig. 15, to illustrate the effect of inhibitory sites within an entire dendritic tree on the control of the generation of an impulse by an EPSP spreading to the soma.

A general conclusion from this latter type of analysis is that inhibitory sites become increasingly effective vis-à-vis control over the cell body as they are placed closer to the cell body. But this is subject to many qualifications. Timing, for example, is a crucial factor. In the model of Fig. 16, a steady IPSP was postulated for the sake of simplicity, but if we assume instead a brief IPSP, then, obviously, a well-timed IPSP at the middle dendrite will be more effective than an inappropriately timed IPSP at the soma. Also, Rall has made computations for more complicated branching systems and has shown that an inhibitory location that is identical with a peripheral excitatory location can produce more effective inhibition than an equal amount of inhibition applied at the soma.

This should be sufficient to indicate that many factors enter into the relative siting of excitatory and inhibitory synapses vis-à-vis the cell body and axon hillock. The nervous system provides examples of virtually every possible variation on this theme. In the stretch receptor cell of the crayfish, for example, inhibitory input has a peripheral location, near the site of excitatory input from the receptor terminals. Motoneurons provide examples of overlapping excitatory and inhibitory inputs throughout the dendritic tree (Chapter 5). Hippocampal pyramidal neurons have inhibitory input directed to their cell bodies (Chapter 11); olfactory and neocortical cells also have inhibitory synapses on the initial segments of their axons (Chapters 10 and 12). These differences are obviously important to an understanding of the synaptic organization of different regions of the brain.

In addition to passive spread, a synaptic response may lead to the generation of an impulse. There are several possible relations between synaptic sites and active sites. There may be a patch of active membrane near the input site in the dendrites, which, in effect, serves as a booster for further electrotonic spread through the dendritic tree. There may be active membrane in the intervening dendrites, through which an impulse can propagate. And there may be active membrane only at a distance, with passive spread thereto. We will encounter examples of all these situations. For this reason, the study of dendritic electrotonus is a part of the larger study of *dendritic properties*, and this will, therefore, be our larger focus in the study of the neurons of each region.

In conclusion, it may be admitted that an understanding of dendritic electrotonus requires an effort that goes beyond the descriptions of dendritic branching patterns and the observations of physiological responses. But the extent to which these mathematical methods can be applied to the interpretation of the anatomical and physiological results is perhaps a direct measure of our progress toward building the foundations for a science of synaptic organization. Toward this end, an intuitive grasp of the methods is far better than none. The need has been succinctly expressed by Katz (1966):

> It can fairly be said that the cable properties of the neuron are the physical basis for all integrative processes at central nervous synapses. For example, spatial summation (or subtraction) of synaptic effects that interact within an effector cell depend upon the spread of subthreshold electric signals along the cell membrane. And as the synapses are clustered closely together within a fraction of a millimeter of the cell body, local integration of such signals can be handled by the subthreshold cable properties of the membrane. . . . It seems that the core conductor, or cable mechanism, is not involved in the transmission of signals *across* most synapses, although it undoubtedly is the basis of the *integration* of local messages, once these have been transferred to the common effector cell.

Part 2

ORGANIZATION OF NEURONAL SYSTEMS

5

SPINAL CORD: VENTRAL HORN

The control of motor behavior is perhaps the most obvious and characteristic function of the central nervous system. Many parts of the brain are involved in this control; directly or indirectly, they all ultimately influence the brain stem and spinal cord, wherein lie the cells—the motoneurons—whose axons connect to the muscles. Within the cord, the motoneuron cell bodies are located in the ventral (anterior) horn, as is indicated in the diagram of Fig. 17. In Sherrington's felicitous phrase, the motoneuron is the "final common path" for the control of movement.

We have already seen that the motoneuron receives sensory inputs from the periphery through the dorsal root ganglion (DRG) cell. This was indicated in the diagram of Fig. 1. The fact that the DRG cell axons lie in the periphery has been of inestimable value to the experimenter; the axons can be stimulated by electrical shocks, or their receptors can be selectively activated by natural stimuli. If we add to this the fact that the axons of the motoneurons also lie in the periphery, and that they connect to muscles whose activity can be easily observed and measured, it can be readily appreciated that one has a near-ideal situation for an experimental analysis of input-output relations. This has afforded a shortcut, as it were, to the analysis of the spinal cord. From this approach, exploited so brilliantly by Sherrington, have developed the concepts of the reflex arc and the integrative or-

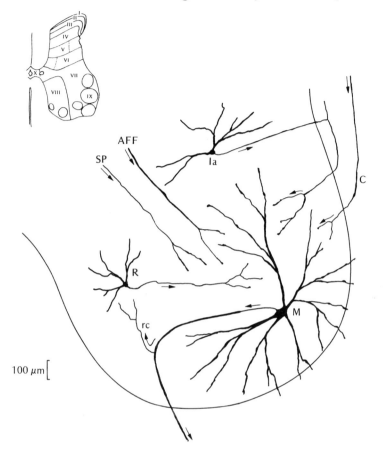

FIG. 17. Neuronal elements of the ventral horn of the mammalian spinal cord.

Inputs: central (C) descending from brain; relays from other parts of spinal cord (SP); afferent sensory fibers from the periphery (AFF).

Principal neuron: motoneuron (M) with recurrent collateral (rc).

Intrinsic neurons: interneuron in Ia afferent pathway (Ia); interneuron in Renshaw pathway (R).

Bracket, 100 μm; the drawings of neuronal elements in other chapters are to the same scale.

ganization of these arcs at the level of the spinal segment, concepts familiar to every student of biology, physiology, and psychology.

In recent years, there has been a growing realization that the spinal segmental apparatus must be viewed as part of its central, as well as its peripheral, connections. This realization has been abetted by improved techniques for studying activation of motoneurons through central pathways. From this work, it is possible, for the first time, to begin to gain a perspective on the integrative context of the motoneuron relative to internally generated behavior (which, after all, is what interests us most) as well as to reflex responses (see Burke, 1971; Evarts, 1971). Our reconstruction of the synaptic organization of the ventral horn will reflect this new orientation.

NEURONAL ELEMENTS

The spinal cord is a very complex structure. As is well known, the central gray matter is in the form of dorsal and ventral horns, as seen in a cross section. A division of the gray matter into laminae was proposed by Rexed (1954), and his system of terminology has come into general use; the inset in Fig. 17 provides a summary diagram. The region occupied by the motoneurons is lamina IX (actually, it is more in the form of a nucleus) and will be the focus of our interest.

The first step in understanding the organization of the ventral horn is to identify clearly the neuronal elements contained therein. We will proceed systematically in a sequence that will be the same for all regions of our study. First to be identified are the long-distance axons that bring input to the ventral horn. Next are the axons that carry output from the ventral horn; in identifying these, we at the same time specify and describe the principal neuron that gives rise to these axons. Last, we identify the intrinsic neurons whose processes and connections are contained entirely within this region.

INPUTS The inputs to the ventral horn fall into two main categories, peripheral and central. The *peripheral* or *afferent* fibers are, as already indicated, the axons of the DRG cells. The diameters of these fibers cover a wide range, from 20μm down to about 2μm. Most of the smaller axons are unmyelinated.

A wide range of sensory modalities is carried in this array. The classical concept has been that different modalities are carried in different-sized axons; from muscle stretch receptors in the largest axons; from touch and pressure receptors in the large- and intermediate-sized axons; and from pain and temperature receptors in the finest axons. Many of these borders are becoming blurred, however, from the results of recent work; for example, pain and temperature, as well as some tactile, stimuli are carried in axons of different diameters.

As Fig. 1 indicates very schematically, the afferent fibers, upon entering the spinal cord, join one of a number of tracts that ascend to higher centers in the brain. In addition, they give off branches within the spinal cord, the primary afferent collaterals. Some of these collaterals reach the ventral horn directly, as shown in Fig. 17. Here they bifurcate and run in a longitudinal direction for one to four segments, giving off terminals in their course. By virtue of these branches, there is *divergence* of the peripheral input; by virtue of the overlap of branches from different axons, there is *convergence* of the peripheral input. These are fundamental features in characterizing the inputs to all local regions in the brain.

Other collaterals of the DRG axons terminate elsewhere (e.g., the intermediate nucleus), from whence there are relays to the ventral horn as well as to other regions.

Central inputs are classified not by their fiber diameter or their "modality" but by their site of origin within the brain. The most distant site is the cerebral cortex. As indicated in Fig. 1, there is, in many mammals, a direct connection from the cortex to the spinal cord. These fibers form the *corticospinal tract.* As shown in Fig. 17, these fibers (labeled C) distribute directly, in primates, to the ventral horn and indirectly, through connections in the intermediate gray (labeled SP). This pathway from the cortex is called *suprasegmental,* to distinguish it from the pathways within and between the spinal cord segments themselves. There are other suprasegmental pathways descending from other parts of the brain, for example, the *red nucleus,* the *reticular nucleus,* and the *vestibular nucleus.* These are not shown in Fig. 17 for the sake of simplicity. Like the peripheral input axons, these central input fibers characteristically distribute longitudinally over several spinal segments.

There are also many pathways within the spinal cord itself, connecting one segment with another (*intersegmental*) and one side with the other (*commissural*). Although these pathways are essential to the mechanisms of the spinal cord, we will concentrate on the other connections. The same applies to the commissural pathways of the other regions we will study.

PRINCIPAL NEURON   The output of the ventral horn is to the muscles, and the principal neuron whose axon carries this output is the *motoneuron*. There are two types of motoneurons, depending on the muscle fibers innervated. The *alpha motoneuron* sends its axon to skeletal (extrafusal) muscles; it is the main type of principal neuron. The *gamma motoneuron* sends its axon to the intrafusal fibers of muscle spindles. These fibers make a negligible contribution to the work done by the skeletal musculature, but their contractile state (controlled by the activity of the gamma motoneurons) sets the sensitivity of the spindle as it monitors the extrafusal muscle contractions. In mammals, the gamma motoneurons are specialized for this particular function. Such a clearcut division among principal neurons is unusual for local regions in the brain.

The alpha motoneurons have cell bodies that range in size from 25-100 μm in diameter; they are, thus, among the largest neurons in the vertebrate brain, as well as among the most variable in size. Each cell body gives rise to several large dendritic trunks, 10-15 μm in diameter at their origin. These are among the largest dendrites in the brain. The dendrites branch relatively modestly and reach considerable lengths, up to 1000 μm (1 mm) distance from the cell body. The dendritic surface is generally smooth; there are occasional spiny protuberances, and varicosities may be present near the terminations.

The gamma motoneurons are, on average, smaller than the

alpha motoneurons, with cell bodies ranging from 15-40 μm in diameter. Their dendritic trees are, by and large, smaller versions of those of the alpha motoneurons. The size difference applies also to the axons, the alpha axons being 12-20 μm in diameter, the gamma axons 2-9 μm. The designations alpha and gamma, in fact, derive from the classical terminology for the size ranges of the peripheral axons.

Within the ventral horn, the motoneurons have a definite topographical arrangement. The motoneurons that supply the muscles of the trunk and neck are located medially, whereas those that supply the muscles of the limbs are located laterally. The motoneurons for the limbs are further localized to the cervical and lumbar enlargements of the cord, which supply the upper and lower limbs, respectively. The motoneurons that supply the more distal limb muscles, particularly those of the primate hand that are under the finest control, are located more dorsally and more caudally than those that supply the proximal muscles. These aspects are admirably described by Brodal (1969).

We say, then, that there is a *somatotopical organization* of the ventral horn, reflecting, to some extent, the spatial relations of the muscles innervated. The groupings of motoneurons take the form, not of rounded nuclei, but, rather, of longitudinal columns within the ventral horn, a fact not adequately brought out by the cross-sectional diagram of Fig. 17. The motoneuron dendrites, like the input axons mentioned above, also reflect this by their predominantly longitudinal orientation; there is considerable overlap of the dendritic trees of the motoneurons within these longitudinal columns. Because the ventral horn is spread out over a considerable longitudinal extent in this way (40 cm, i.e. 400,000 μm) in the adult human), it is much more difficult to define "local region" here than in most other parts of the brain.

Near its origin from the cell body, the motoneuron axon gives off one or several collateral branches. Like the dendrites, these have a generally longitudinal orientation. They attain lengths of 1-2 mm and are believed to terminate in the ventral horn of the same spinal segments (Ryall, Piercey, and Polosa, 1971). Some motoneurons do not have axon collaterals, at least none that are impregnated by the Golgi method. We will see that axon collaterals are characteristic of the principal neurons in most (but not all) of the regions of the brain. Motoneuron axon collaterals are considered to be intrinsic processes, confined within the region from which they arise. They are termed *recurrent collaterals*, to distinguish them from collateral branches that only extend laterally, and from collateral branches that distribute to different regions (as, for example, the primary afferent collaterals of the DRG axon).

In the periphery the axons of motoneurons branch and terminate on the muscles to form there the presynaptic processes of the neuromuscular junctions. A single alpha motoneuron, together with all the muscle fibers it innervates, was termed a *motor unit* by Sherrington. The size of a motor unit (i.e., the number of axonal branches to different muscle fibers, or divergence ratio) is a measure of the fineness of motor control; the smaller the unit, the finer the control. The variation in motor unit size is quite considerable, from about 100 for the smallest muscles of the hand to 1900 for the large muscles of the leg. If we consider the different kinds of muscles in the body—the extraocular muscles that control the eye, the muscles in the larynx, and the muscles in the face, as well as those in the hands, limbs, and trunk—it should be evident that we are dealing with systems that may be quite different in many respects. This is a necessary caution in drawing inferences about general principles of motor organization that apply to all these systems.

INTRINSIC NEURONS There has never been any doubt that the ventral horn contains neurons other than motoneurons; they are usually termed *propriospinal neurons* by anatomists and *interneurons* by electrophysiologists. Their functional identity, however, has been very difficult to establish. Only very recently has it been possible to stain cells subjected to electrophysiological analysis (Jankowska and Lindström, 1971, 1972) and identify thereby the types of intrinsic neurons that are involved in particular pathways.

The cell bodies of intrinsic neurons are scattered throughout the ventral horn, but, from recent studies, it appears that there are two main regions of concentration. One is situated just dorsal to lamina IX (see Fig. 17). The cell bodies here have diameters of 20-30 μm. Each cell body gives rise to several dendritic trunks, 3-5 μm in diameter. The dendrites have generally smooth surfaces, they branch sparingly, and they terminate some 200-500 μm from the cell body. They, thus, resemble the smaller motoneurons in size and dendritic pattern. It is noteworthy that the dendrites have a dorsal-ventral orientation, in contrast to the longitudinal orientation of motoneuron dendrites. These cells have been identified as the interneurons in the Ia inhibitory pathway to the motoneurons (Jankowska and Lindström, 1971; see Fig. 19 and later).

The other area of concentration of intrinsic neurons is near the site where the motoneuron axons gather to leave the ventral horn. The cells stained in this region by Jankowska and Lindström (1972) are, in general, similar in size and shape to those just described. The evidence that these cells may be the interneurons in the Renshaw pathway for recurrent inhibition of the motoneurons will be discussed later in the section on synaptic actions.

These two types of intrinsic neuron fall into the category of propriospinal neuron. The general similarity in form between intrinsic neurons and motoneurons is a notable feature of the ventral horn. In most of the regions of the brain in our study, the intrinsic neurons can be sharply differentiated from the principal neurons in size and shape. It is important, in this regard, to differentiate between the propriospinal neuron and the type known as a *short-axon cell*. In many regions of the brain we will find cells of this type, the axon of which is very limited in its extent and distribution. Although definite short-axon cells are present in other parts of the spinal cord, they appear to be scarce or absent in the ventral horn (Schiebel and Schiebel, 1966; Jankowska and Lindström, 1971, 1972).

The term propriospinal denotes the fact that the axon distributes within the ventral horn itself. This is the basis upon which we designate it as an intrinsic neuron. The extent of the axon varies considerably, however; some axons distribute within one or two segments of their cell of origin, while others distribute many segments away. The axon characteristically enters a tract (see Fig. 17) through which it runs to the distant sites. It, therefore, becomes a matter of definition whether these propriospinal axons are to be regarded as intrinsic neurons, or as another type of principal neuron; this is yet another example of the problem of determining what constitutes a local region within the spinal cord. A similar problem is encountered with regard to the types of neurons in the cerebral cortex that give rise to association fibers (see Chapters 10 through 12).

NEURONAL POPULATIONS An analysis of synaptic organization must ultimately include quantitative data about the number of neuronal elements involved. We will, therefore, note such data as are available. They are, unfortunately, very difficult to obtain, and they almost invariably represent rough estimates, so rough, in some cases, as to be misleading. An additional problem is that populations vary greatly in different species. This is dramatically shown for the spinal cord by the comparative study summarized in Table 1. The great increase in numbers in this series reflects both an increase in size and an increase in complexity of motor control as one ascends the vertebrate series. One moral of this example is that quantitative data have little meaning unless the

Table 1

Quantitative estimates of numbers of peripheral afferent input fibers (in dorsal root) and of output axons from motoneurons (in ventral root) for spinal cords of different species

	DORSAL ROOT (*afferent axons*)	VENTRAL ROOT (*motoneurons*)	RATIO
Toad	8,000	5,700	1.4:1
Mouse	45,000	23,000	1.9:1
Dog	300,000	150,000	2:1
Man	1,000,000	200,000	5:1

From Agduhr (1934).

species from which they were obtained is identified. The student may well ponder the implications of this fact in considering the possibility of general models of ventral horn organization.

In the human spinal cord, the total number of motoneurons on one side has been estimated at about 200,000, as shown in Table 1. The ratio of alpha to gamma motoneurons is about 3:1, and one obtains, therefore, a total of some 150,000 alpha motoneurons and 50,000 gamma motoneurons. These totals are quite modest compared to the numbers of principal neurons in other regions of the brain; compare, for example, the 7,000,000 Purkinje cells in one side of the human cerebellum (Chapter 8). The average number of motoneurons per spinal segment works out to be about 5500 alpha motoneurons; this is so small as to raise the question of whether a single segment can be considered an independent local region.

Two quantitative ratios are of fundamental importance in the study of synaptic organization. One is the convergence ratio of input fibers to principal neurons. The data in Table 1 can only indicate an upper limit for this ratio as it applies to the ventral horn, since many of the dorsal root fibers go elsewhere in the cord. The ratios, at any rate, can be seen to vary from 1.4 in the toad to about 5 in the human. Thus, this ratio increases to a much smaller degree than the numbers of neuronal elements. This ratio for the spinal cord is very low compared to that for many other regions of the brain; in the retina, for example, the over-all ratio may be as high as 1000:1 (see Chapter 7). Individual afferent fibers, however, diverge through multiple branches and terminals, as has been described above. This increases the over-all convergence ratio, although at the expense of spatial discreteness in the input-output relations.

The other ratio of importance concerns the relative number of intrinsic and principal neurons. In anatomical studies, the ratio of small neurons (presumably intrinsic neurons) to motoneurons, in the ventral horn of the cat, has been estimated to be about 7:1 (see Brodal, 1969). This is greater than in some other regions (e.g., the thalamus) but much smaller than in others (e.g., the olfactory bulb). In most of the regions of the brain, the number of intrinsic neurons reflects their relative contribution to the processing of information in the input pathways and in the pathways for lateral and recurrent control. As we shall see, this is not so obvious in the ventral horn.

SYNAPTIC CONNECTIONS

Although the ventral horn has certain topographical patterns of organization, as described above, it lacks the rigid separation of neuronal elements into laminae that is so characteristic of other parts of the brain in our study. Because of this fact, identification of the synaptic connections between the neuronal elements is a much more difficult task than in other regions.

There are two major morphological types of synaptic terminals in the ventral horn of mammals. As is illustrated in Fig. 18(A), one type, the larger, is 1-4 μm in diameter. The synapse consists of a presynaptic cluster of spheroidal vesicles and a prominent densification of the postsynaptic membrane. These are, therefore, type I chemical synapses, as described in Chapter 2. These synapses are found at all levels of the motoneuron—the cell body and the proximal, intermediate, and distal dendrites—although the distribution tends to favor the distal dendrites.

The other major kind of synaptic terminal is somewhat smaller, being 0.5-3 μm in diameter [Fig. 17(B)]. The vesicles tend to be flattened or variable in shape, and the synaptic membranes are more equally dense. These, therefore, tend to fall into the category of type II chemical synapses. They, too, are found at all levels of the motoneuron, but their distribution tends to be skewed toward the cell body. Each of these types accounts for approximately 40% of the total population of terminals in the ventral horn, so together they provide most of the input to the motoneurons.

Among the other much less frequent types of synaptic terminals, one is of particular interest. This is the largest type in the ventral horn, with a diameter 4-7 μm. The contacts made by

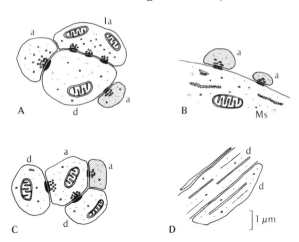

FIG. 18. Synaptic connections in the mammalian ventral horn. (A) Axodendritic synapses from axon terminals (a) to a motoneuron dendrite (d). Note the multiple contacts made by the Ia axon terminal. (B) Axosomatic synapses onto a motoneuron cell body (Ms). (C) Synaptic cluster with serial synapses (axoaxodendritic). (D) Dendrodendritic membrane juxtaposition.

Terminals that make type II chemical synapses are shaded in this and corresponding diagrams in other chapters. Bracket, 1 μm; the drawings of synaptic connections in other chapters are to same scale.

(After Bodian, 1966, 1972; Charlton and Gray, 1966; Uchizono, 1966; Conradi, 1969; and McLaughlin, 1972a, b; Matthews, Willis, and Williams, 1971.)

these terminals are type I chemical synapses. One of these single terminals may have as many as five to eight synapses onto a single postsynaptic profile (McLaughlin, 1972b). Multiple contacts from one terminal to another are unusual in the nervous system, although we will encounter an even more striking example of this type of arrangement in the retina (Chapter 7).

The identification of the sources of the different types of terminals is a forbidding task, and there is relatively little evidence with regard to most of the type I and type II synapses described above. The best evidence relates to the large terminals. After dorsal root section, these terminals degenerate, and it is inferred that they are the monosynaptic terminals from the largest fibers in the dorsal root, the Ia afferent axons from the muscle spindles. There is general agreement that these terminals constitute less than 1% of the total number onto motoneurons.

From the preceding description, it can be seen that most of the connections in the mammalian ventral horn are made by simple terminals with simple synapses, as defined in Chapter 2. The main orientation of the synapses is axosomatic and axodendritic. In these respects, the synaptic repertoire of the ventral horn is rather limited; lacking are the specialized terminals and the rich variety of interconnections, particularly between dendrites, that are found in such other regions as the olfactory bulb, retina, and thalamus. Nor is there the grouping of terminals into synaptic complexes and glomeruli, as is seen in the cerebellum and thalamus.

Irregular small groups of terminals are sometimes seen, however, and of interest in this respect is the occasional serial sequence of synapses, as is illustrated in Fig. 17(C). A large terminal can be seen making a synapse onto two presumed dendritic branches; in addition, the terminal is postsynaptic to another terminal. It has been postulated (Gray, 1962) that such sequences provide the morphological basis for presynaptic inhibition of motoneurons. It is not clear, however, whether such sequences are present within the lamina IX (McLaughlin, 1972b) or whether they are in nearby parts of the spinal cord. Similar sequences have been reported in the dorsal horn (see Ralston, 1968), where it is possible that some of these sequences are dendrodendritic rather than axoaxonal.

It may finally be noted that there are membrane-to-membrane appositions between motoneuron dendrites in some species [see Fig. 17(D)]. These appositions may provide an opportunity for the spread of current between the motoneurons. No tight or gap junctions have been reported in mammals, although they have been found in the frog (Sotelo and Taxi, 1970).

The number of synapses on a single motoneuron has been estimated to be as high as 20,000 (Gelfan, 1963). This is one of the largest numbers for any neuron in the brain, exceeded only by cortical pyramidal cells and cerebellar Purkinje cells. The total surface area of cell body and dendrites of a large motoneuron can be calculated to be as high as 100,000 μm^2, so it appears that there is room for this many synapses. Most of the motoneuronal surface area (estimated 80%) and most of the synapses are on the dendrites (Aitken and Bridger, 1961).

BASIC CIRCUIT

It will be convenient to summarize the main patterns of synaptic connections for each of the local regions we study in a diagram. Such a diagram may be referred to as a *basic circuit*. It will serve as a useful reference in several respects: for discussing the principles of organization of that region; for subsequently describing synaptic actions and dendritic properties; and for comparing with the organization of other regions.

In the ventral horn, the lack of distinct cell types and lamination makes the task of identifying the main synaptic pathways from the available anatomical evidence alone very difficult. For that reason, the basic circuit depends to a great extent on the results of the electrophysiological analyses of synaptic transmission. We anticipate that evidence in the diagram of Fig. 19.

A prominent feature of the ventral horn is the large number of different functional pathways that feed into it. Figure 19 is a very schematic diagram that summarizes only some of the simplest of these functional pathways. Particular emphasis is given to the central connections of the motoneurons and to the organization of the interneurons. The orientation is somewhat different from that of the usual diagrams of spinal cord organization in that the central pathways (1-4) are here considered as the main inputs to the ventral horn and arrive from the top of the diagram. The peripheral afferents (5 and 6) are considered as feedback inputs from the sites of ventral horn output and arrive, therefore, from the bottom of the diagram.

Let us first consider the central input pathways. Most central motor control in vertebrates is effected through fibers descending

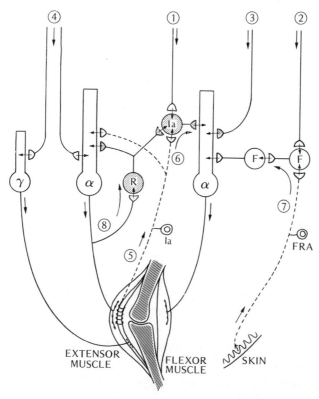

FIG. 19. Basic circuit diagram for the mammalian ventral horn. *Input pathways:* central descending pathways for the brain (1-4) (see text); the Ia afferent excitatory pathway from muscle spindles (5); the Ia afferent inhibitory pathway (6); the flexor afferent pathway (FRA) from the skin (7); the Renshaw recurrent inhibitory pathway (8). *Principal neurons:* alpha (α) motoneurons to flexor and extensor muscles; gamma (γ) motoneurons to intrafusal muscle fibers of muscle spindles. *Intrinsic neurons:* Ia interneuron in the Ia afferent pathway; F interneuron in the flexor reflex pathway; R, Renshaw cell in the recurrent inhibitory pathway. Neurons and terminals with presumed excitatory actions are in open profiles; those with presumed inhibitory actions are shaded in this and corresponding diagrams in other chapters.

from the brain that relay through interneurons in the spinal cord. These relays may be through either inhibitory (pathway 1) or excitatory (pathway 2) interneurons. Depending on the number of synaptic relays, these pathways are either *disynaptic* or *poly-synaptic*. The interneurons in these pathways may be the ones located within the ventral horn itself (i.e., the Ia interneuron in pathway 1), or they may be located in other parts of the spinal cord (i.e., the intermediate nucleus of Cajal and the dorsal horn). *Monosynaptic* pathways from the brain stem and cerebral cortex provide direct central control of the motoneurons. These pathways play onto both flexor (pathway 3) and extensor (pathway 4) motoneurons. The central monosynaptic pathways are especially directed to motoneurons that supply such distal muscles as those of the hand (see Phillips, 1971). Pathway 4 in Fig. 19 indicates that central activation of the alpha motoneurons to the hand muscles is frequently linked to activation of the gamma motoneurons that supply the spindles in the hand muscles. Thus, at the same time that the alpha motoneurons cause the muscle (an extensor muscle, in the example of Fig. 19) to contract and shorten, the gamma motoneurons cause the spindles of that muscle to contract and shorten also. This was called *alpha-gamma linkage* by Granit (1955) and is considered to be the mechanism whereby the length of the muscle spindle is adjusted according to the length of the surrounding muscle, so that the spindle continues to be sensitive to changes in length or tension as movements are carried out.

The orientation of the diagram in Fig. 19 emphasizes this role of the muscle spindle in providing feedback information to the motoneuron from the site of motoneuron output. This information travels, as an impulse discharge, in the large Ia axons of DRG cells (pathway 5), and it has two destinations within the ventral horn. One destination is an excitatory monosynaptic connection onto motoneurons that supply that muscle and its synergists. The other destination (pathway 6) is a disynaptic connection through an inhibitory interneuron (the Ia interneuron in Fig. 19) onto motoneurons that supply antagonist muscles. This is an expression of the Sherringtonian principle of *reciprocal inhibition*, which is considered to be basic to the reciprocal relation of muscle movements about a joint. Note that these Ia inhibitory interneurons are the same ones that are under central control from descending pathways, i.e., the reciprocal organization is built into the central as well as the peripheral connections. This important new concept is discussed more fully by Lundberg (1970) and Burke (1971).

Muscle spindles, thus, have a role to play within the whole context of centrally generated movements, in addition to their more local role in reflex behavior at the spinal level. As is well known, this latter role revolves around the *stretch reflex*. Traditionally, this has been regarded as the basic spinal mechanism for the maintenance of muscle tone and the activation of extensor muscles that oppose gravity and thereby maintain upright posture.

In addition to the Ia afferents from muscle spindles, there are also Ib afferents from Golgi tendon organs and II afferents from the spindles. These afferents have their own spinal connections that contribute to centrally generated movements and to particular types of reflex behavior. For simplicity, they are omitted from Fig. 19. Muscle receptors and their central actions are fully described by Matthews (1972).

In addition to muscle afferents, there are many other types of peripheral inputs. For the most part, they make polysynaptic excitatory connections, particularly onto flexor motoneurons (pathway 7), and are, therefore, often referred to collectively as *flexor reflex afferents* (FRA). As is well known, they provide the basic mechanism for reflex withdrawal of a limb from a noxious or painful stimulus. The FRA interneurons are located in other parts of the spinal cord; like the Ia interneurons, they are also part of central descending pathways.

In addition to the long feedback loops through the periphery, there are also short feedback loops within the ventral horn due to the presence of motoneuron axon collaterals. As indicated in pathway 8, the axon collaterals connect to an interneuron, the so-called Renshaw cell, that makes inhibitory connections onto motoneurons as well as interneurons. This is the pathway for recurrent inhibition in the ventral horn. Note that the Renshaw cell also connects to Ia interneurons, which, themselves, are inhibitory to motoneurons. Through this connection, the recurrent inhibitory pathway can bring about a decrease of the feedforward inhibition through the Ia pathway, an action that is termed *disinhibition*. There are many permutations of these kinds of effects within the local circuits of the ventral horn.

In later chapters, we will have frequent occasion to compare aspects of the organization of other brain regions with those of the ventral horn. For the present, it will suffice to note certain of the most important aspects. First, the pathways in the ventral horn are mainly from axons to cell bodies and dendrites; there is a conspicuous absence of local interactions between dendrites. In this respect, the ventral horn resembles the cerebellum (Chapter 8) and differs from the olfactory bulb (Chapter 6), retina (Chapter 7), and thalamus (Chapter 9). Second, the triad of input, output, and intrinsic elements is diffusely arranged; this is similar to most parts of the cerebral cortex (Chapters 10-12) and differs from the tightly organized triadic relation between these elements in the olfactory bulb, retina, cerebellum, and thalamus. Third and finally, most of the input pathways onto the motoneurons are polysynaptic, through interneurons in the ventral horn and elsewhere in the spinal cord. The excitatory interneurons provide for a preliminary integration and processing of inputs to the motoneurons; we will see examples of this kind of relay in the granule cells of the cerebellum, the granule cells of the dentate fascia (Chapter 11), and, possibly, the stellate (granule) cells of the neocortex as well (Chapter 12). Inhibitory interneurons also provide for a functional inversion, or commutation, of excitatory to inhibitory inputs; among the other regions of our study, the neocortex offers possible examples of this type of relay.

SYNAPTIC ACTIONS

Analysis of synaptic actions within the ventral horn must begin with the simplest and most accessible input to the motoneurons. This is through the largest Ia fibers from the muscle spindles (pathway 5 in the basic circuit diagram of Fig. 19). The electro-

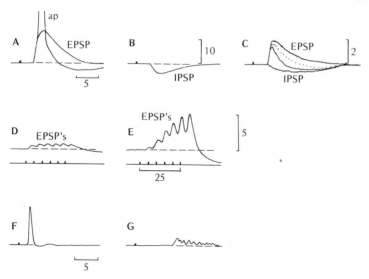

FIG. 20. Main types of synaptic actions in the ventral horn, as recorded intracellularly from the motoneuron. (A) EPSP at threshold for generating an action potential (ap) (After Eccles, 1953.) (B) IPSP. (After Eccles, 1957.) (C) Integration of EPSP and IPSP gives the result shown in the middle trace, which differs from a simple algebraic summation (dotted line). (After Rall et al., 1967.) (D) Reproducible EPSP's elicited over the Ia monosynaptic pathway. (E) Rapid buildup of EPSP's elicited over the monosynaptic corticospinal pathway. (From Phillips and Porter, 1964.) (F) Extracellular recording of a brief discharge in motoneuron axons (ventral root) in response to a single Ia volley (pathway 5 in Fig. 18). (G) Prolonged response to a single volley in flexor reflex afferents (pathway 7 in Fig. 18). (After Lloyd, 1943.) Vertical bars, millivolt calibration; horizontal bars, milliseconds, in this and corresponding diagrams in other chapters.

physiologist proceeds by delivering a single shock to a dorsal root or muscle nerve in order to set up a synchronous volley of impulses in these fibers. The impulses invade the axon terminals and activate the synapses onto the motoneurons. The pioneering experiments of Eccles and his co-workers (Eccles, 1953) showed that an intracellular electrode within a motoneuron records a depolarizing response, such as that in Fig. 20(A). The latency of onset of the response is brief but sufficient to allow for a 0.5-msec synaptic delay, as is characteristic of chemical synapses. This is, therefore, a monosynaptic EPSP, mediated by a type I chemical synapse. It has a relatively brief time course: a rapid rise of several milliseconds and a longer decay of 10-20 msec.

The threshold for generation of an action potential is about 15 mV. Electrophysiological analysis has shown that the impulse initiation site is in the axon hillock and initial segment. The spread of the synaptic potential to the impulse initiation site is governed by the electrotonic properties of the motoneuron dendrites. This will be explained further in the next section.

The volley in the Ia spindle afferents also activates the interneurons that connect to antagonist motoneurons (pathway 6 in Fig. 19). From such a motoneuron, the hyperpolarizing response shown in Fig. 20(B) is recorded. This is an IPSP, presumably mediated by the type II chemical synapses in the ventral horn. These IPSP's have simple brief rise times and decays, the time course being somewhat longer than that for the monosynaptic EPSP's.

The interaction of an EPSP and an IPSP serves as a paradigm for synaptic integration as discussed in Chapter 3. Figure 20(C) shows recordings from an experiment in a motoneuron, in which an EPSP (upper trace) and an IPSP (lower trace) were timed to occur simultaneously. The middle trace shows the result of this interaction; it is significantly different from the dotted line, which indicates the transient that would have occurred if the interaction had been an algebraic summation of the two. In some cases, the interactions are, in fact, linear, but in other cases, as in this example, the interaction is nonlinear. This illustrates the point made in Chapter 3 that the interactions between EPSP's and IPSP's depend on the interactions of the synaptic conductances and the point made in Chapter 4 that they also depend on the relative positions of the synapses and their electrotonic relations within the dendritic tree.

The response of a motoneuron to a sequence of three Ia volleys is shown in Fig. 20(D). The later responses resemble the first; there is a dead-beat quality to the sequence. But synapses do not all have this simple action, as has been shown in experiments on the central monosynaptic inputs to the motoneurons. One such input comes through the pyramidal tract fibers that originate in the cerebral cortex (cf. Fig. 1). Figure 20(E) shows that, when this pathway is repetitively stimulated, there is a dramatic increase in amplitude of the synaptic potentials (Phillips and Porter, 1964). This increase is due to an overlap of the synaptic potentials (*temporal summation*) as well as to the increased amplitudes of successive responses (*facilitation*).

These findings are significant for several reasons, as Phillips and Porter have pointed out. They show that central synapses differ in their potency. They also show that the high frequency discharges of which pyramidal neurons are known to be capable would be especially effective in activating the motoneurons over the monosynaptic pathway. This pathway appears to provide for the fractionation of movement that underlies the exquisite control of the hand. The directness of this pathway should increase the accessibility of hand motor units to the complex intracortical neuronal systems that control the cortical pyramidal output. Last, the increased synaptic potency is what is needed for the cerebral cortex to override other inputs to the motoneuron in its role of initiating and controlling sudden or precise movements. We will later discuss the even more dramatic changes in synaptic potency in the hippocampus (Chapter 11) and neocortex (Chapter 12).

These monosynaptic responses may be compared with the responses of motoneurons to inputs arriving over polysynaptic pathways. As shown in Fig. 20(F, G), the response to even a single volley in the flexor afferent pathway (FRA, pathway 7, in Fig. 19) consists of a response that builds up slowly and long outlasts the initial volley. Many factors may be involved in this type of response: delayed transmission through the synaptic relays in the spinal interneurons; prolonged transmitter action and prolonged EPSP's; prolonged spike discharges aroused by the EPSP's; and reverberating activity due to re-excitation loops among the circuits of the spinal cord. Similar activity can be induced by stimulation of central pathways. It is clear that much

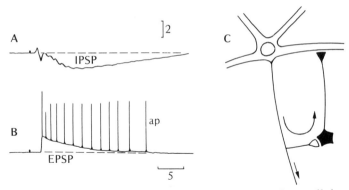

FIG. 21. Recurrent inhibition in the ventral horn. (A) Intracellular recording of a prolonged IPSP in a motoneuron elicited by a volley in motoneuron axons (ventral root). (B) Intracellular recording of a high frequency discharge of a Renshaw unit. Postulated pathway for activation of the Renshaw cell and inhibition of the motoneuron. (After Eccles, 1969 and Willis, 1970.)

of the ongoing control of motoneurons is mediated by this type of activity, but it is equally clear that analysis of these pathways is one of the most difficult challenges facing the electrophysiologist.

RECURRENT INHIBITION   Among the local circuits of the ventral horn, the recurrent inhibitory pathway has been the most intensively studied and has been of the greatest interest as a model for similar pathways in other regions of the brain. It has already been shown in Fig. 19 (pathway 8). The inhibitory action of this pathway is best revealed by setting up a single volley in a ventral root that invades the motoneurons antidromically. The response in a motoneuron consists of the antidromic impulse followed by a prolonged hyperpolarization, as shown in Fig. 21(A). By using a shock that excites only the axons of neighboring motoneurons, it can be shown that the hyperpolarization is not the afterpolarization of the impulse but, rather, is due to an IPSP. The latency of the IPSP is sufficient for two synaptic relays, and it is concluded that the pathway is from the motoneuron axons through their collaterals onto an interneuron that has inhibitory synapses onto the motoneurons.

The name proposed by Eccles, Fatt, and Koketsu (1954) for this interneuron is the Renshaw cell, in honor of Birdsey Renshaw, who obtained the first physiological evidence for its existence in 1946. The cells respond to the ventral root volley with a prolonged discharge of impulses, as shown in Fig. 21(B). The frequency is extremely high, starting at 1500/sec (i.e., an interspike interval of only about 0.6 msec); the spikes are, correspondingly, extremely brief in time course and refractory period. The cells are difficult to record from, apparently being rather small. The intracellular recordings show that the spikes arise from an EPSP with an unusually sharp initial peak and a very long tail; the latency of onset is about one synaptic delay before the onset of the motoneuron EPSP. The frequency of the discharge is similar to the frequency of the ripples that are sometimes discerned on the IPSP in the motoneuron; see Fig. 21(A).

These various lines of evidence have indicated that the single volley of impulses in motoneuron axon collaterals sets up a prolonged EPSP and spike discharge in these interneurons, which, in turn, set up the prolonged IPSP in motoneurons through inhibitory synapses from their axon terminals [Fig. 21(C)]. There is strong evidence that acetylcholine is the transmitter substance from the synapses of the collaterals onto the interneurons; this is consistent with the fact that acetylcholine is also the transmitter at the other output from the axons at the neuromuscular junction. This is often cited as an example of the workings of Dale's Law (Chapter 3). The transmitter substance for the inhibitory synapses onto motoneurons is as yet undetermined.

This then is the Renshaw pathway in the spinal cord. It was the only model for recurrent inhibitory pathways in the central nervous system until the discovery of the dendrodendritic pathway in the olfactory bulb, to be described in the next chapter. A nagging problem, however, has been the histological identification of the Renshaw cell itself. It was early supposed that it was a type of short-axon cell, but, as we have seen, this type is rare in the ventral horn. The possibility that Renshaw discharges are recorded from the terminal boutons of the motoneuron recurrent collaterals themselves has been suggested by Weight (1968) and supported by intracellular-marking experiments (Erulkar, Nichols, Popp, and Koelle, 1968). The recent experiments of Jankowska and Lindström (1971) support the anatomical evidence that the Renshaw cell is a type of propriospinal neuron. Because of these uncertainties, many observers prefer to use the noncommittal term "Renshaw element".

In addition to the question of identification, the role of recurrent inhibition in the ventral horn has been by no means clear. An analogy with surround or contrast inhibition in sensory systems has seemed obvious, but the evidence for such a function has not been compelling. Renshaw cells may be involved in complex disinhibitory interactions with each other (Wilson and Burgess, 1962) and with the interneurons in the Ia inhibitory pathway onto the motoneurons (Hultborn, Jankowska and Lindström, 1968). The unusual properties of these elements—their abruptly rising and long-lasting EPSP's and their extremely high discharge rates—are not understood. All these unanswered questions indicate that there is still much work to be done on the significance of recurrent inhibitory pathways in the ventral horn (see Willis, 1971, for a comprehensive review).

DENDRITIC PROPERTIES

Like roads leading to Rome, the pathways in the ventral horn lead to the motoneuron. The terminals are mainly distributed over the dendrites, so the study of input-output mechanisms in the ventral horn comes down, largely, to a study of the properties of the motoneuronal dendrites. This was the original object of Rall's analysis, and modern studies of synaptic integration in the motoneuron are now firmly based on his methods. The motoneuron has thus come to serve as a model for the analysis of the dendritic properties of other neurons in the brain.

The basic approach involves the construction of an equivalent cylinder for the dendritic tree of the motoneuron. The steps in this procedure have been outlined in Chapter 4 and are summar-

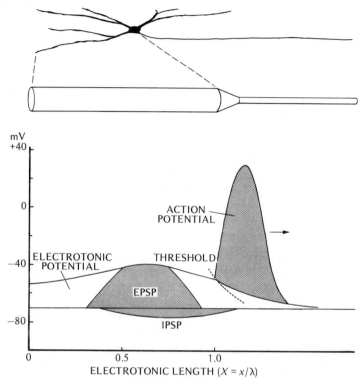

FIG. 22. An electrotonic model of a motoneuron, illustrating the spread of potentials in the dendritic tree. *Above,* derivation of an equivalent cylinder from the motoneuron. *Below,* types of potentials and their distribution (see text). *Ordinate,* intracellular potential (mV). *Abscissa,* electrotonic length.

ized in the diagram of Fig. 22. The electrical properties used in constructing the cylinder are derived from experiments in which current is passed through an intracellular electrode, while the voltage change is simultaneously recorded. Such experiments have yielded values between 0.5 and 5 $M\Omega$ (average, 1.65 $M\Omega$) for the whole neuron resistance (R_N) of the motoneuron (Frank and Fuortes, 1956) and a time constant for the charging transient of 2-3 msec. From such data, estimates of 4-7 msec for the membrane time constant (τ_m) and 2000-4000 Ωcm^2 for the specific membrane resistance (R_m) are obtained (Rall, 1959b, 1970b; Lux et al., 1970). These experiments, combined with measurements of motoneuron dendritic geometry, have permitted estimates of the electrotonic length (L) of the equivalent cylinder. The best estimates appear to lie between 1 and 2 (Burke and ten Bruggencate, 1971; Jack and Redman, 1971); a value of 1 is used in the illustration in Fig. 22. In recent experiments, it has been possible to carry out these studies in single motoneurons identified with intracellular stains, by autoradiographic techniques (Lux et al., 1970), and with Procion dye (Barrett and Crill, 1974).

In our discussion of dendritic electrotonus, it was shown that the shape of a synaptic potential, as recorded from the cell body, varies according to its site in the dendritic tree because of the electrotonic properties of the dendrites (Chapter 4; see Fig. 15). This principle was early recognized in the motoneuron by Fadiga and Brookhart (1960). As developed by Rall, the shape indices of synaptic potentials have provided a useful tool for ascertaining the sites of some of the synaptic inputs to the motoneurons. The case of the Ia monosynaptic input is illustrated in Fig. 22. The moderately rapid rise time and decay of the EPSP indicate that the synapses are widely distributed over the motoneuron surface, with a predominance in the middle portion of the dendritic tree. This is indicated diagrammatically by the shaded area in Fig. 22.

The synaptic potential must spread to the site of impulse initiation in the initial segment, as previously mentioned, and Fig. 22 shows how this comes about. The motoneuron dendritic membrane is electrically passive, hence, the spread of synaptic potentials is governed by the electrotonic properties of the dendrites. The significance of the electrotonic length calculated for the dendritic tree is that it is sufficiently short to allow for effective spread of passive current through the dendrites to the cell body. Thus, the synaptic potential in the dendrites is linked to the action potential in the initial segment by the electrotonic potential in between.

This is a common arrangement in neurons. For example, a

model for the stretch receptor cell of the crayfish would be very similar to that shown in Fig. 22, except that the location of the excitatory input would be more distal in the dendrites. In that cell, too, the site of impulse initiation is well out in the initial segment, as first shown in the classical experiment of Edwards and Ottoson (1957). This illustrates an important principle, namely, that the location of the cell body has no necessary significance in the integrative organization of a neuron. In the motoneuron, it happens to be (usually, but not always) at the convergence point for current flow from dendrites to initial segment. In other neurons in the brain, we will see examples in which the cell body is completely irrelevant for particular input-output relations.

Motoneurons of different size have been found to have similar electrotonic lengths for their equivalent cylinders; the values for L lie between 1 and 2 (Burke and ten Bruggencate, 1971). Thus, in the smaller motoneurons, the smaller diameters of the dendrites are balanced by their shorter lengths. This introduces a *scaling principle* in the electrotonic properties of dendritic trees of different size, which, as we will see, may have a general application to neurons in other parts of the brain (cf. Rall and Rinzel, 1973).

The electrotonic properties of the smaller motoneurons (and of smaller neurons, in general) have an important consequence: a given synaptic input gives rise to a larger synaptic potential in them. This is because the resistance (the *input resistance*) through which the synaptic current will have to flow in the smaller dendrites is higher. This may be related to the fact that smaller motoneurons tend to be more excitable and display more spontaneous activity. The alpha motoneurons have low levels of spontaneous activity (0-10 impulses/sec), whereas the smaller gamma motoneurons fire at rates of 30-50/sec. The functional significance of this difference is that the continual activity of the gamma motoneurons keeps the spindles slightly contracted, thereby providing continual Ia feedback excitation of the alpha motoneurons, whose resulting slow discharges are the basis for resting muscle tone. Such differences in spontaneous activity between different types of neurons are important factors in understanding the dynamics of synaptic organization, as we will also see in other regions of the brain.

Differences in excitability between alpha motoneurons of different size have been found; this has been termed a *size principle* by Henneman (1966). According to this principle, increasing activity over a given input pathway successively excites motoneurons according to their increasing size. The extent to which this principle, deduced from electrophysiological experiments, is operative in the normal ongoing control of the muscles is currently under investigation. With regard to other parts of the brain, the principle is consistent with what is known about the activity of neurons in the olfactory bulb (Chapter 6) but not in the cerebellum (Chapter 8).

In the electrotonic model for the motoneuron, it is assumed that the spread of the synaptic potentials through the dendritic tree is by passive means alone. This appears to be generally true for the normal motoneuron. In motoneurons undergoing chromatolysis after sectioning of their axons, however, small amplitude spike-like activity can be recorded, and it has been concluded that this indicates the generation of impulses or impulse-like activity in restricted parts of the dendritic tree (Eccles, Libet, and Young, 1957). This is an important finding; it provides a model for active properties of dendrites in other parts of the brain (see the cerebellar Purkinje cell (Chapter 8) and the hippocampal pyramidal cell (Chapter 11), and it shows that the properties of dendritic membrane depend, to a certain extent, on the functional and metabolic state of the cell.

SYNAPTIC INTEGRATION   The locus for Renshaw recurrent inhibition is also shown in Fig. 22. It is mainly in the proximal dendrites, closer to the cell body than the Ia excitatory input. In the early experiments it was inferred that this reflected a general rule, that inhibitory synapses are placed close to the cell body in order to be most strategically placed to control impulse generation. As discussed in Chapter 4, this rule no longer has a general validity. In the crustacean stretch receptor cell, for example, the inhibitory

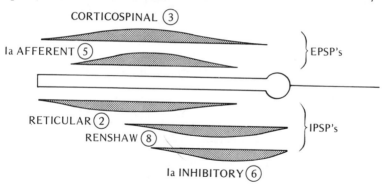

CORTICOSPINAL ③

Ia AFFERENT ⑤ } EPSP's

RETICULAR ②
RENSHAW ⑧ } IPSP's

Ia INHIBITORY ⑥

FIG. 23. Distribution of some synaptic inputs to the motoneuron. Numbers refer to pathways in Fig. 19. (After Porter and Hore, 1969; Burke et al., 1970; Kuno and Llinás, 1971; Jack, Miller, Porter, and Redman, 1971.)

synapses occur far out on the dendrites. The strategy of that placement appears to be to permit them to directly oppose the excitatory sensory input to the dendritic terminals. In the motoneuron there appears to be a mixture of the two cases: some inhibitory synapses directly oppose the excitatory synapses through the overlap in their distribution (in the distal dendrites, this results in remote inhibition; see below). Others tend to be closer to the cell body, the more effectively to control impulse initiation in the initial segment. These two kinds of inhibition are very clearly seen in the cerebellar Purkinje cell (Chapter 8).

The distributions of several of the other inputs to the motoneuron are shown in Fig. 23. Some of these distributions have been deduced by Rall's methods, others from such evidence as susceptibility to intracellular Cl^- and reversal by intracellular currents. In some cases, different parts of a motoneuron dendritic tree may receive different inputs; this was termed a "fractionation of dendritic field" by Sprague (1958). In general, however, there are relatively wide distributions and extensive degrees of overlap between the sites of central, afferent, and intrinsic connections. These are important features in the synaptic organization of motoneurons and are in contrast to the organization of many other principal neurons of the brain, in which there is a sharp localization of different inputs to different parts of the dendritic tree. The overlap of inputs provides the fullest means by which the many different inputs to the motoneuron can interact dynamically with each other.

To understand the nature of these interactions, it is necessary to know what the input provided by a single synapse is. The large amplitude EPSP, depicted in Fig. 22, represents the summated response to many individual synapses. For a single Ia afferent terminal, Kuno (1971) has obtained evidence that a single quantum (i.e., possibly the amount of transmitter contained in a single synaptic vesicle) causes a unit ("miniature") EPSP of about 100 μV and that there are, on average, only 1-2 quanta/ terminal. Therefore, if the impulse threshold is about 10 mV, the synchronous input from about 50 Ia terminals is needed to initiate an impulse in the motoneuron. It may be noted that the amplitude of a postsynaptic potential depends on input resistance and spatial distribution; depending on these variables, the amplitude of unit synaptic potentials elicited by other inputs may vary from 0.05 to 0.5 mV.

We can now see more fully the significance of the fractionation of inputs by means of the small contacts from small terminals, as discussed in Chapter 2. It means that a single input is rarely effective itself in initiating or controlling output from the motoneuron; it must always combine with many others. The summing of the inputs must depend on accurate timing and on appropriate siting vis-à-vis the electrotonic properties of the dendritic tree. Because each input is small, the summation will be finely graded. And because the impulse threshold is relatively high, a great amount of finely graded interaction occurs in the dendrites before output occurs in the axon. Since there are no dendrodendritic synapses, there are no opportunities for local outputs from the dendrites, as we shall see in certain other neurons. The ongoing control of the muscles is, therefore, the outcome of a complex process in the motoneuron, in which a multitude of local synaptic interactions precedes the final integration at the site of impulse initiation in the axon hillock.

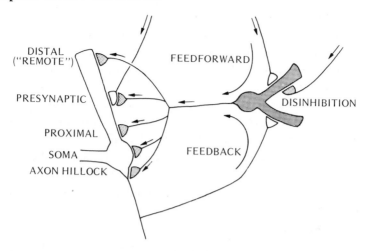

FIG 24. Types of local inhibitory connections and pathways onto moto-neurons.

TYPES OF INHIBITION   It should be evident by now that many different types of inhibition are involved in mechanisms of synaptic integration, and it is appropriate to distinguish between the types as they affect the motoneuron. Inhibition directed predominantly to control of impulse output is indicated at the *soma* and *axon hillock* in Fig. 24. In motoneurons of lower vertebrates (e.g., Mauthner cells in the goldfish), this type of control is quite specific, through highly differentiated terminals that surround the hillock region. This type of inhibition tends to prevent impulse initiation without spatial discrimination of the sources of excitatory inputs. We will see examples of this type of control in the cerebellum (Chapter 9) and hippocampus (Chapter 11).

Most synaptic inputs to mammalian motoneurons are directed to the dendrites; a *proximal* inhibitory synapse is indicated in Fig. 24. There are also many terminals on the more *distal* dendrites. Because their action may not be evident in intracellular recordings taken from the cell body, it is often referred to as *remote inhibition*. We have already pointed out that its significance lies mainly in its local interactions with other dendritic inputs.

A final type is the so-called *presynaptic inhibition,* in which inhibition is exerted over an excitatory terminal. Through this type of connection, the excitatory input can be effectively opposed without affecting the excitability of the motoneuron itself; it is, therefore, hard to distinguish it experimentally from remote inhibition. This type has been demonstrated in electrophysiological experiments, and there is electron-microscopic evidence for synaptic sequences between axon terminals [see Fig. 18(C)], but its role during natural activity has not yet been determined. We will see that presynaptic interactions between dendritic terminals are important in the organization of the olfactory bulb, retina, and thalamus.

The foregoing types of inhibition are defined in relation to the motoneuron itself; indeed, their significance cannot be assessed without reference to the electrotonic properties of the motoneuron. In addition, there are inhibitory actions that are defined in relation to intrinsic pathways. The two main types are depicted in Fig. 24. *Feedback inhibition* has already been discussed in relation to the Renshaw pathway. *Feedforward inhibition,* on the other hand, is exemplified by an interneuron in an input pathway, as shown in Fig. 24. There has been much interest recently in the role of feedforward circuits, both excitatory and inhibitory, in the central control of movement (see Evarts, 1971). Finally, Fig. 24 shows an inhibitory connection to the inhibitory interneuron, a connection providing for *disinhibition* of the inhibitory input from the interneuron to the motoneuron.

All these types of pathways are found generally throughout the brain, and we will see many examples in the regions of our study. The terminology has been explained here in relation to inhibitory pathways, but many of the terms apply as well to excitatory pathways. In the ventral horn, the Renshaw interneurons and the Ia interneurons have rather different distributions; other spinal interneurons are localized in the intermediate nucleus and the dorsal horn. In other regions, we will see examples of extreme segregation of certain types of interneurons for specific types of feedforward and feedback actions.

6

OLFACTORY BULB

The olfactory bulb, protruding like an incandescent fixture from the forebrain, is the main relay station in the olfactory pathway. It occupies in this respect a position similar to that of the retina in the visual pathway. It is here, as is shown in the diagram of Fig. 25, that the axons of the olfactory receptor cells in the nose terminate. The output of the bulb is directed to several brain regions, and there are, in return, connections to the bulb from the brain. The main outlines of these connections are indicated in Fig. 25; they will be described in further detail below and in the later chapters on the olfactory cortex and hippocampus.

The classical histologists of the late nineteenth century took a keen interest in the olfactory bulb. This was by virtue of its very distinct laminations and its several sharply differentiated types of neurons. The deductions drawn by Cajal (1911, 1955) and his contemporaries about the organization of neurons in the olfactory bulb played a central role in the development of the functional concepts of the classical neuron doctrine. Following this period, however, interest waned, due in large part to the almost complete lack of progress in understanding the nature of the olfactory stimulus.

Interest in the olfactory bulb was revived by the anatomical studies of le Gros Clark and his co-workers (Clark, 1951; 1957; Allison, 1953) and the physiological studies of Adrian (1950;

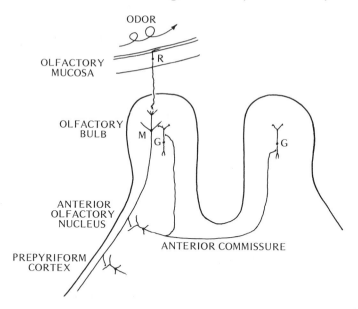

FIG. 25. Connections of the olfactory bulb. R, olfactory receptor cell; M, mitral cell; G, granule cell.

1953). Knowledge of synaptic organization began in the early 1960's, with electrophysiological investigations using single unit recordings. These studies took advantage of the fact that the input and output pathways of the bulb are completely separated, the kind of situation that made the ventral horn of the spinal cord such an admirable subject for study. Electron-microscopic and biophysical studies followed, which have together put our knowledge on a sound basis.

The results of these studies have shown that, in its functional organization, the olfactory bulb goes significantly beyond the framework of the classical neuron doctrine as formulated on the motoneuron model. It is appropriate, therefore, to consider the olfactory bulb at this point, in order to compare it with the ventral horn and to use it as a bridge to two other highly organized regions of the brain, the retina and cerebellum.

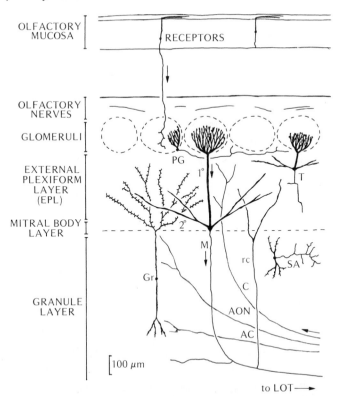

FIG. 26. Neuronal elements of the mammalian olfactory bulb.

Inputs: afferent fibers (above) from olfactory receptors; central fibers (below) from three sources; centrifugal fibers (C) from the nucleus of the horizontal limb of the diagonal band; ipsilateral fibers from the anterior olfactory nucleus (AON); contralateral fibers from the anterior commissure (AC).

Principal neurons: mitral cell (M), with primary (1°) and secondary dendrites (2°) and recurrent axon collaterals (rc); tufted cell (T).

Intrinsic neurons: periglomerular short-axon cell (PG); deep short-axon cell (SA); granule cell (Gr). LOT, lateral olfactory tract.

NEURONAL ELEMENTS

INPUTS   The neuronal elements of the olfactory bulb are shown in Fig. 26. The *afferent* (peripheral) input is through the axons of the receptor cells in the olfactory mucosa in the nasal cavity.

The axons are all unmyelinated and extremely thin, approximately 0.1-0.3 μm in diameter. The fact that the input fibers are all of the same modality (olfactory), and they all have small diameters and no myelin, stands in the sharpest possible contrast to the variety of input fibers to the ventral horn of the spinal cord.

The olfactory axons enter at the bulb surface and terminate in a layer composed of spherical regions of neuropil, termed *glomeruli*. These structures are 100-200 μm in diameter; like the glomeruli in the cerebral cortex (see Chapter 12), they are at a macroscopic level of organization compared with the synaptic glomeruli of the cerebellum and thalamus. A general definition of the term glomerulus, as it applies to different levels of organization, is given in Chapter 8.

The olfactory axons do not branch on their way to the glomeruli, but once inside they ramify, to varying extents, and terminate. An essential feature of bulbar organization in all vertebrates is that all the olfactory input terminates within the glomeruli.

There are several *central* inputs to the bulb from the brain. Their sites of origin have been indicated in the diagram of Fig. 25. Axons that are relatively large (several microns in diameter) but few in number come the farthest distance, from a region at the base of the brain called the diagonal band (DB). Other axons, finer and much more numerous, come from the region just posterior to the bulb, the anterior olfactory nucleus (AON). Some of these come from the AON of the same side; others come from the contralateral side through the anterior commissure (AC). The central inputs are also referred to as *centrifugal* fibers to indicate their outward orientation from the brain.

These inputs have terminals at different though overlapping levels in the bulb, which is indicated very approximately in the diagram of Fig. 26 and will be further described later. It is another essential feature of bulbar organization that the centrifugal fibers are directed mainly to the deeper layers of the bulb.

PRINCIPAL NEURON   The output from the bulb is directed centrally (not peripherally, as in the case of the ventral horn) and is carried in the axons of *mitral cells*. As indicated in Fig. 26, the mitral cell bodies are arranged in a thin sheet about 400 μm below the glomerular layer. As principal neurons go, the mitral cell is small-to-medium sized, its cell body being 15-30 μm in diameter. Each cell sends an unbranched *primary dendrite* to a glomerulus, to terminate there in a tuft of branches. The tuft has an extent of 100-150 μm and, therefore, fills much of the glomerulus it lies within. The diameter of the primary dendrite ranges from 2-12 μm, and the length is 400-600 μm, depending on the angle of direction across the external plexiform layer (EPL). Each mitral cell also gives rise to several *secondary dendrites*, which branch sparingly and terminate in the EPL. Secondary dendrites are 1-8 μm in diameter and up to 600 μm or so in length.

The primary and secondary dendrites have generally smooth surfaces, like the dendrites of motoneurons. They are also similar to motoneuron dendrites in diameter and length. But the differentiation into primary and secondary types, the strict localization of afferent input to the terminal tuft of the primary dendrite, and the separation of intrinsic circuits in relation to the two types of dendrite, are specializations that have no counterparts in the motoneuron.

The mitral cell axons range from 0.5-3 μm in diameter, the larger ones being myelinated. They proceed to the depths of the bulb and then run posteriorly to emerge together to form the lateral olfactory tract (LOT). During their course within the bulb they give off two kinds of collaterals: *recurrent collaterals* that terminate in the EPL and *deep collaterals* that terminate in the granule layer (GRL).

Smaller versions of the mitral cells, the so-called *tufted cells*, have cell bodies scattered throughout the EPL. Many of these appear to send axons to the LOT and, thus, carry part of the bulbar output. They are, therefore, a smaller type of principal neuron. Whether they are differentiated to perform some specific function, as are the smaller gamma motoneurons of the ventral horn, is not known. Their collaterals appear to be preferentially distributed within a thin layer just deep to the mitral cell bodies.

The output axons in the LOT distribute collateral branches and terminals to several central regions; these are indicated in Fig. 25 and will be further discussed in Chapter 10. The distances over which this output is carried are rather short, not much more than a centimeter or so. Note that the principal neuron of the olfactory bulb terminates on several distinct types of distant region, each with different functions; this is in contrast to the motoneuron, whose output goes exclusively to one destination, the muscles. Thus, the two regions offer an interesting contrast in that the input is multimodal and the output is unimodal in the case of the motoneuron, but the reverse in the case of the mitral cell.

INTRINSIC NEURONS   In contrast to the difficulty in identifying interneurons in the spinal cord, we can recognize three distinct types of intrinsic neuron in the olfactory bulb.

Surrounding the glomeruli are the cell bodies of *periglomerular* (PG) *cells* (see Fig. 26). The cell body is small (6-8 μm in diameter), far smaller than any cell in the ventral horn, and among the smallest in the brain. Each cell has a short bushy dendritic tree that arborizes to an extent of 50-100 μm within one of the glomeruli. The axon distributes to neighboring glomeruli, reaching distances of 500 μm or so. Significantly, the axon does not distribute to the glomerulus containing the dendritic tree of its parent cell.

We have previously touched on the identification of *short-axon cells* in discussing the intrinsic neurons of the ventral horn, and we may now deal more fully with this question with regard to the PG cells. We will define a short-axon cell as *a cell whose axon distributes within the same histologically homogeneous region.* Note that this differs slightly from the oft-quoted classical definition, of a cell whose axon ramifies within the field of the dendritic tree of its parent cell. Thus, once we have characterized a region in terms of its characteristic nuclear or laminar arrangements, any cell whose axon distributes within that region is a short-axon cell. A cell whose axon distributes outside the local region is consequently a long-axon cell and, by definition, a principal neuron.

In the case of "loosely organized" parts of the brain, such as the ventral horn and, as we shall see, the cerebral cortex, a clear distinction between the two neuronal types may sometimes not be possible. In the case of more rigidly organized regions, however, the distinction is usually clear. Within this general definition of a short-axon cell, we will encounter many varieties: axons that distribute within the dendritic field of the parent cell (e.g., some cells in the thalamus) and those that distribute outside it (e.g., the PG cell); those axons that distribute within the same histological lamina (e.g., the PG cell) and those that distribute within a different lamina (e.g., some cells in the cerebellum).

Deep to the layer of mitral cell bodies is a thick layer containing the cell bodies of *granule cells* (see Fig. 26). These cell bodies are also very small (6-8 μm in diameter); hence the name granule, a very general term applied by the early histologists to the "grains" they saw in their primitive microscopic preparations (see also Chapter 8). Each granule cell has a superficial process that ramifies and terminates in the EPL. The branching tree has a lateral extent of 300-500 μm within the EPL. The shaft of the tree is up to 500 μm or more long, depending on the depth of the cell body from which it arises. In their branching pattern and their investiture with many small spines (gemmules), the superficial processes resemble the apical dendrites of cortical pyramidal cells. They contrast with the smooth appearance and infrequent branching of the dendrites of mitral cells and of motoneurons. Each granule cell also gives off an inner process that terminates deeper in the granule layer.

The outstanding feature of the granule cell is that it lacks a morphological axon; repeated Golgi studies and, more recently, electron-microscopic studies, have confirmed this fact. In this respect the granule cell resembles the amacrine cell of the retina. These cells have thus always stood out as exceptions to the classical model of the neuron based on the motoneuron. The lack of an axon has raised the question of the identity of the processes of the granule cell; some would call them "primitive" or "axon-like". One need not hesitate, however, in calling them dendrites on the grounds of their general fine structural features and close resemblance to cortical cell dendrites. The lack of an axon has posed a severe challenge to the interpretation of the functions of these cells, which has only been accomplished in the studies of recent years.

There is also within the bulb a third type of intrinsic neuron, a *short-axon cell* found rarely in the glomerular layer but more frequently in the granule layer. The latter have cell bodies 8-15 μm in diameter and dendritic trees of variable extent (up to 200-300 μm across). Their axons ramify either in the EPL or the GRL (see Fig. 26).

CELL POPULATIONS   The numbers of receptor cells have been estimated at about 50,000,000 on one side of the nose in the rabbit; this is therefore the number of afferent axons entering one olfactory bulb. This number of afferent elements is exceeded only by the number in the visual pathway (see Chapter 7). The number of mitral cells is of the order of 50,000, so there is considerable convergence, on the order of 1000:1, onto the principal neurons. There are about 2000 glomeruli in a rabbit's bulb, so there are approximately 25,000 olfactory axons and 25 mitral cells per glomerulus (Allison and Warwick, 1949).

The ratios of intrinsic neurons to principal neurons are also high in the olfactory bulb. Approximate estimates are 20:1 for the PG cells:mitral cells, 200:1 for the granule cells:mitral cells, and perhaps 1:1 for the short-axon cells : mitral cells. These ratios by themselves strongly support the presumption that a considerable amount of information processing takes place within the intrinsic circuits of the olfactory bulb.

SYNAPTIC CONNECTIONS

The severe lamination of the olfactory bulb greatly simplifies the analysis of synaptic connections as seen under the electron mi-

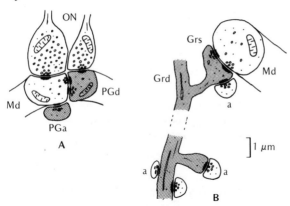

FIG. 27. Synaptic connections in the mammalian olfactory bulb. (A) Axodendritic and dendrodendritic connections in the olfactory glomerulus. ON, olfactory nerve; Md, mitral dendrite; PGd, periglomerular cell dendrite; PGa, periglomerular cell axon. Note the serial and reciprocal synaptic sequences. (After Andres, 1964; Reese and Brightman, 1970; Pinching and Powell, 1971; White, 1972.) (B) *Above*, dendrodendritic connections in the external plexiform layer between a mitral secondary dendrite (Md) and a granule cell dendritic spine (Gr s); note also the centrifugal axodendritic connection onto the spine. *Below*, axodendritic connections in the granule layer. (After Rall et al., 1966; Price and Powell, 1970a.)

croscope; in addition, identification has been put on a sound basis by the extensive use of serial reconstructions. The analysis has been carried out at three main levels in the bulb.

GLOMERULAR LAYER   Within the glomeruli the terminals of olfactory axons make synaptic contacts onto the dendritic tufts of the principal neurons: the mitral (and tufted) cells (Andres, 1970; Reese and Brightman, 1970). As shown diagrammatically in Fig. 27, the terminals are moderately large (1.0-3.0 μm in diameter), particularly when viewed in relation to the very thin axons from which they arise. The contacts are type I chemical synapses. The axon terminals also make synapses onto the dendrites of the intrinsic neurons, the PG cells, in most species that have been studied. An exception is a particular strain of mouse (Balb/c), in which these connections are missing. This finding has been well documented by White (1972, 1973) in serial reconstructions and points up a very interesting genetic difference at the level of synaptic connectivity.

By the use of serial reconstructions and careful comparisons with Golgi preparations, it has been established that there are numerous synaptic connections between the mitral and PG cell dendrites within the glomerulus (Pinching and Powell, 1971; White, 1972). The mitral-to-PG synapses are type I, whereas the PG-to-mitral synapses are type II [see Fig. 27(A)]. These dendrodendritic synapses may be arranged in serial sequences or in reciprocal, side-by-side pairs (cf. Fig. 6, Chapter 2). About 25% of the synapses are involved in reciprocal pairs. A group of axonal and dendritic terminals is sometimes set apart to form a synaptic complex in a manner somewhat resembling the synaptic glomeruli in the cerebellum and thalamus. Synapses between two presumed PG cell dendrites have also been observed. At the borders of the glomeruli are axon terminals on the larger mitral dendritic branches; these terminals come from PG cell axons; a few also come from central inputs (DB).

EXTERNAL PLEXIFORM LAYER (EPL) In the EPL, the dominant type of synaptic connection is a pair of reciprocal contacts (Hirata, 1964; Andres, 1965; Rall, Shepherd, Reese, and Brightman, 1966). Serial reconstructions (Rall et al., 1966) established that these contacts occur between the secondary dendrite of a mitral cell and the spine (gemmule) of a granule cell dendrite [see Fig. 27(B)]. These were the first dendrodendritic synapses identified in the nervous system. The mitral-to-granule synapse is type I, whereas the granule-to-mitral synapse is type II (Price and Powell, 1970a). Over 80% of all synapses in the EPL are involved in such reciprocal pairs. If we consider that there are a hundred or more granule cells for each mitral cell, and that each granule cell has perhaps a hundred or more spines, it is obvious that these dendrodendritic connections provide for extremely powerful and specific interactions with the mitral cells. Indeed, electron micrographs show the EPL to be a neuropil composed almost entirely of mitral and granule cell dendrites and their synaptic interconnections (see Reese and Shepherd, 1972).

GRANULE LAYER   In the granule layer, axon terminals are found on the shafts and spines of the granule cell dendrites [see Fig. 27(C)]. The studies of Price and Powell (1970b) have shown that these axon terminals derive from both intrinsic and extrinsic (central) inputs. The *intrinsic* sources include the axon collaterals of mitral and tufted cells and the axons of the deep short-axon cells. There is evidence that the synapses of these terminals are type II. The *extrinsic* sources have been shown to make connections at different levels of the granule dendritic tree. The AC distributes mainly to the deep processes. The AON distributes over the middle part of the dendrites, including the spines in the EPL. The DB axons distribute mainly to the spines in the EPL; some terminals are also found at the borders of the glomeruli. The synapses made by these inputs from the brain appear to be type I. In the EPL, the synapses made by these terminals are on the same spines that take part in the reciprocal connections with mitral dendrites, providing thereby a means for presynaptic control of the spines.

It should be noted that all the synaptic connections in which the granule cell takes part are oriented toward the granule cell, with the sole exception of the dendrodendritic synapses from the granule spines onto the mitral dendrites in the EPL. The latter are, therefore, the only output avenue from the granule cells.

It can be seen that there is a much richer variety of patterns of synaptic connections in the olfactory bulb than in the ventral horn. In the ventral horn, nearly all the connections are axodendritic or axosomatic, with the exception of some serial arrangements presumed to be axoaxonic. The olfactory bulb also contains many axodendritic synapses, but, in addition, there are dendrodendritic, somatodendritic, dendrosomatic, and axodendrodendritic synapses, as described above. The purely local circuits are, therefore, much more complex in the olfactory bulb than in the ventral horn.

GLIA The olfactory bulb has provided favorable opportunities for observing the relation of glial membranes to the neuronal elements and synapses. The incoming olfactory axons are gathered into bundles of 100-200 axons contained within one glial (Schwann) cell. Within the glomeruli, a synaptic complex, such as that shown in Fig. 27(A), is often seen to be surrounded by one or more loose folds of glial membrane; it has been noted that this is similar to, though not nearly as distinct as, the synaptic glomeruli of the cerebellum and thalamus (Pinching and Powell, 1971). A similar relation is sometimes seen around the reciprocal synapses in the EPL.

Within the EPL, in most vertebrate species, several loose folds of glial membrane surround the primary dendrite of the mitral cell near the glomerular boundary. In primates, it has been found that the folds of membrane are packed down into typical myelin, which may surround not only the primary dendrite but even extend to the cell body in the case of tufted cells (Pinching, 1971). As already noted in Chapter 2, this finding shows that a dendrite may be myelinated and that myelin is not exclusively associated with axons.

BASIC CIRCUIT

The synaptic organization of the olfactory bulb may be summarized in a basic circuit diagram; see Fig. 28. Because of the clear cell types and laminae in the bulb, the diagram can be constructed largely on the basis of the foregoing anatomical evidence.

First, by way of brief summary: the olfactory axons make synapses within the glomeruli onto mitral (and tufted) cell dendrites, as well as onto PG cell dendrites in most species. Between the dendrites of mitral and PG cells are synaptic connections, both reciprocal and serial. The PG cell axons connect to mitral dendrites in neighboring glomeruli. In the EPL, the major type of connection is the reciprocal dendrodendritic synapse between mitral dendrites and granule dendritic spines. Finally, the input fibers from the brain have terminals at several levels on the gran-

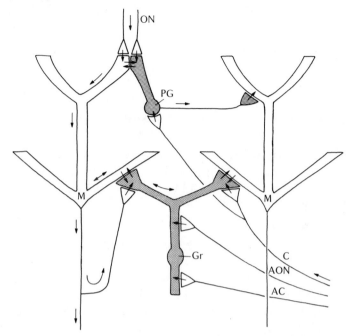

FIG. 28. Basic circuit diagram for the mammalian olfactory bulb. Abbreviations as for Fig. 27.

ule cells: AC axons to the deep granule processes; AON axons to the middle portions; and DB axons to the peripheral dendritic spines.

With regard to the three main types of neuron, an outstanding feature is the fact that they all take part in dendrodendritic synaptic connections. The principal neuron (mitral cell) not only has synaptic *inputs* to all parts of its dendritic tree, it also has synaptic *outputs* from virtually every part of that dendritic tree. Much the same can be said of the PG cell dendrites. The granule cell also has synaptic inputs over all its dendritic tree, but its synaptic outputs are localized to the spines in the EPL. Thus, the granule cell differs from the other two types of bulbar neuron not in having synaptic output from its dendrites (all three types have that) but rather in having that output localized to only one part of the tree, and in lacking an additional axonal output pathway.

The diagram of Fig. 27 illustrates the principles upon which this vast variety of interconnections is organized. An obvious feature of the basic circuit is that there are vertical pathways for direct, "straight-through" activity and horizontal pathways for interactions between the vertical paths. The horizontal connections are organized in two distinct levels, or tiers. The first level is the glomerular layer, which is obviously concerned with reception of the olfactory input and the initial processing of that input. The second level is the EPL, which is obviously concerned with control of the bulbar output from the mitral cell bodies. The intrinsic neurons are specific to these levels and functions, the PG cells to input processing, the granule cells to output control.

In the general arrangement just described, there is an interesting similarity to the basic circuit of the retina. This will be discussed further in the following chapter. For now it may be noted that in both these regions there is a framework of vertical and horizontal pathways such as cannot be discerned in the ventral horn. Also, in both regions, there is little ambiguity about the positions of the intrinsic neurons in the vertical and horizontal pathways, so that the functional roles of the intrinsic neurons are much clearer than in the case of the ventral horn.

Let us now look more closely at the synaptic organization within the olfactory bulb, using Fig. 28.

The olfactory glomeruli are the most characteristic feature of the olfactory bulb in the vertebrate series. Within the glomeruli, three neuronal elements come together: olfactory axon, mitral cell dendrite, and PG cell dendrite. In more general terms, the three elements are input fiber, principal neuron, and intrinsic neuron. These are the basic elements in the synaptic organization of the local regions of the brain, and as noted (Chapter 1) we will call them a *synaptic triad*. Within the triad, the synapses between the input and principal elements provide the necessary basis for input-output transmission, whereas synapses between the principal and intrinsic elements provide for modulation and control of the input-output transfer.

The synaptic triad in the glomerulus involves synapses from the input onto both the principal and (in most species) the intrinsic elements. This is a common pattern in the brain; the same type of arrangement is found, for example, in the retina, cerebellum, and thalamus. In the latter regions, as we shall see, the synapses onto the principal and intrinsic elements arise from a single large input terminal, whereas, in the olfactory bulb, the synapses are made by separate terminals. The arrangement in the olfactory bulb appears to be more flexible and, possibly, more precise. In order to assess its functional significance, however, one needs to know if the separate terminals arise from separate olfactory axons; if they do, does this mean that some olfactory receptors project only to the principal neuron, whereas others project only to the intrinsic neuron? Such questions are fundamental to understanding the nature of information processing in this sensory pathway.

After the initial input to the principal and intrinsic elements in the glomerulus, further processing takes place through the dendrodendritic synaptic connections between them. In all regions of the brain, the principal and intrinsic elements stand in some kind of relation to each other. In the olfactory glomerulus, the multitude of reciprocal and serial connections appears to maximize the possibilities for complex information processing. A similar variety of connections is present in the retina (see Chapter 7). In the thalamus, the dendrodendritic connections are mainly oriented from the intrinsic to the principal elements (see Chapter 9). In the cerebellum, there is a notable absence of connections between the dendrites of the two elements (see Chapter 8).

These different arrangements will be discussed in due course. Here it may simply be noted that almost every conceivable variation is rung on this triadic theme in the various local regions of the brain; in any given region, the variation is specific for the type of information processing carried out there. The basic patterns are easiest to identify in tightly organized regions, like those just mentioned. In more spread-out regions, like the ventral horn and cerebral cortex, the arrangements are diffuse and more difficult to identify. In these regions, the concept of a synaptic triad seems less applicable, because the connections by a given extrinsic pathway tend to be preferentially directed to either the principal or the intrinsic elements, rather than to both. Even in these cases, however, the three basic elements still stand in some kind of relation to each other, be it in parallel or in sequence, and it is, therefore, useful to use the concept as the basis for understanding the synaptic framework of a region and comparing it with other regions.

If we turn now to the second level of organization in the olfactory bulb, we see that the mitral and granule cell dendrites form the principal and intrinsic elements, respectively. The input elements at this level are of two types. The afferent input comes by way of the mitral primary dendrite; there is therefore an exclusive input to the principal element of the triad by this route. The central input, on the other hand, makes synapses onto the granule cell, and this input therefore is directed exclusively to the intrinsic element.

As a consequence of the two-tier separation of connections within the bulb, the mitral and PG cell synapses within the glomeruli are concerned exclusively with olfactory processing, whereas, at the deeper level, the mitral and granule cell synapses are concerned both with olfactory processing and with integration of information passing forward from the brain through the granule cell, as described above. Some of the information from the brain may be in the form of feedback through long loops from the olfactory projection areas (see Chapter 10). Some of it, however, may be in the form of non-olfactory signals from hypothalamic and limbic structures (Chapters 10 and 11). The granule-to-mitral synapse is, therefore, of interest as a specific site at which there is an overlap of distinct functions. One may characterize it in this regard as a *multifunctional*, or *multiplex*, synapse.

The lamination of central inputs to the granule cell is an important aspect of functional organization, which will be discussed further with respect to dendritic properties.

Two points relevant to the organization of the olfactory bulb as a cortical structure may be noted. From Fig. 28, it can be seen that the initial processing of olfactory input in the glomeruli is exclusively through dendrodendritic interactions, with subsequent processing through interglomerular connections mediated by the PG cell axons. There is a strong implication that a glomerulus may function to some extent as a functional unit. This has been long suspected on anatomical grounds (see Clark, 1957), and there is recent physiological evidence for glomerular specificity for different olfactory stimuli that supports this idea (Leveteau and MacLeod, 1966). It has been pointed out that if glomeruli have this functional specificity, then the group of mitral, tufted, and PG cells with dendrites connected to a particular glomerulus would all share this specificity. This would imply a horizontal constraint on the organization of functionally related neurons in the bulb, which may be analogous to the functional columns of the cerebral cortex (Shepherd, 1972a). We will return to this question when we discuss the glomeruli (barrels) and columns of the cortex (Chapter 12). Recent anatomical studies indicate that there may be several levels of organization within a single olfactory glomerulus (Land, Eager, and Shepherd, 1970).

One of the chief functions of the principal neuron is to provide the vertical conducting and integrating loci across the bulbar cortex, as shown in Fig. 28. The lamination of the bulb, therefore, reflects the sequence of interactions carried out in relation to the principal neuron. A key point is that this sequence is non-repeating: at successive depths, the structures are different, the synaptic connections are different, and the functional interactions are different. There is, therefore, a *non-repeating sequence of structure and function* in relation to the principal neuron at progressive depths in the cortex, and this expresses itself in the sequence of laminae; it is an *oriented multicellular system*. Such considerations as these may be said to touch on the essence of cortical organization, in contrast to the classical considerations

of such points as the numbers of layers, granules, etc. We shall
have occasion to amplify this theme in later chapters.

SYNAPTIC ACTIONS

It would be nice if synaptic actions in the olfactory bulb could
be elucidated using natural stimuli, as can be done in the retina.
But such things are not yet possible, and analysis has had to rely
on electrophysiological techniques. Fortunately, the bulb is admirably suited for this, for it shares with the ventral horn the advantage that its input and output pathways are quite separate and
can, therefore, be activated independently. Since the intrinsic
neurons are small, the analysis of their properties has had to be
based on extracellular recordings of units and summed potentials.

A single volley set up by an electrical shock to the olfactory
nerves gives rise to a sequence of actions depicted on the left of
Fig. 29. The volley reaches the bulb after a long delay (ON in
Fig. 29), due to the slow conduction velocity (0.3-0.5 m/sec) in
the thin unmyelinated axons. The impulse volley depolarizes the
axon terminals, activating excitatory synapses onto the dendrites
in the glomeruli. The EPSP set up in a mitral cell dendritic tuft has
two functions. One is to spread to the primary dendrite and
through it generate impulses in the mitral cell (M in Fig. 29).
The other function is to activate the local synapses of the dendritic tuft onto the dendritic terminals of the PG cells. This
EPSP in the PG cell dendrites also has two functions: it spreads
to the PG cell axon hillock to generate an impulse discharge (PG
in Fig. 29), and, locally, it activates the dendritic synapses of the
PG cell onto other dendrites, including the reciprocal synapse
back onto the mitral cell dendrites. These pathways can be traced
in the basic circuit diagram of Fig. 28. There is evidence that the
output of the PG neuron is inhibitory at both its dendrodendritic
and axodendritic synapses (Shepherd, 1971, 1972a).

We turn next to the actions engendered by the impulse in the
mitral cell body. These have been investigated most directly by
backfiring the mitral cell antidromically from the LOT. An in-

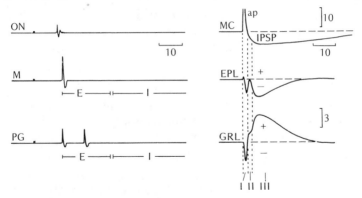

FIG. 29. The main types of synaptic actions in the olfactory bulb.

Left, ON, extracellular recording of an afferent volley in olfactory nerves; M, extracellular recording of the spike response of a mitral cell; PG, extracellular recording of the spike response of a periglomerular cell. Periods of excitatory and inhibitory action, as revealed by a second test volley, are indicated by E and I, respectively. (After Shepherd, 1963, 1971.)

Right, MC, intracellular recording from a mitral cell body, showing an antidromic impulse (ap) followed by an IPSP. (After Yamamoto et al., 1962; Phillips et al., 1963; Nicoll, 1969.) EPL, summed evoked potentials in the external plexiform layer; GRL, summed potentials in the granule layer. Time periods of EPL and GRL response are indicated by dotted lines and I, II, and III below (see text). (After Phillips et al., 1963; Rall and Shepherd, 1968.)

tracellular electrode in the cell body records the response shown in MC in Fig. 29, an impulse followed by a long-lasting hyperpolarization. This has been shown to have the characteristics of an IPSP (Yamamoto, Yamamoto, and Iwame, 1963; Phillips, Powell, and Shepherd, 1963; Nicoll, 1969).

The synaptic pathway for this IPSP was deduced from an analysis of the summed extracellular potentials evoked by the antidromic volley. In the external plexiform layer (EPL), three successive time periods were identified in the evoked potential (Fig. 29). The first two are brief and are related to the invasion of the impulse into the cell body (period I) and the spread into the dendrites (period II); period III, on the other hand, is generated primarily by an EPSP in the granule cell arborization.

The potentials in the granule layer (GRL in Fig. 29) can be divided into the same periods, but they have polarities that are generally the inverse of those in the EPL, and they differ in magnitude, according to a potential divider effect in the external recording circuit. By reconstructing the mitral and granule cell population responses with biophysical models (see below), and incorporating the external recording circuit, it was possible to reconstruct the physiological results. From this analysis it was postulated that the predominant pathway for the recurrent inhibition of mitral cells was by way of dendrodendritic connections through the granule cells. This was confirmed by the electron microscopic evidence, summarized in Fig. 27. The details of the analysis may be found in Rall and Shepherd (1968), Rall (1970), and Shepherd (1972). With regard to the extracellular potential analysis, perhaps the most important point for the general reader to grasp is that it was based on a division of the response into successive time periods, rather than into labeled components, as is the usual practice in studies in other parts of the brain. The stereotyped lamination of the bulb provided an important advantage for this approach.

Certain features of these synaptic actions are of particular interest. One is that the major actions between principal and intrinsic neurons are through dendrodendritic synapses. Enough has already been said to indicate that this in itself should not be considered unusual, if we can free ourselves from the simple view of synaptic organization based on the motoneuron model. We note, in addition, that the dendrodendritic synapses are activated either by a graded presynaptic depolarization (in the case of the synapses in the glomerulus, as well as the granule-to-mitral synapse in the EPL) or by a presynaptic impulse (in the case of the mitral-to-granule synapse in the EPL). The properties of these synapses are characteristic of the neuromuscular junction (Chapter 3) and the axodendritic synapses onto motoneurons (Chapter 5); thus, there is nothing unusual in the mechanism of dendrodendritic synapses, as far as these properties are concerned.

The long-lasting nature of the synaptic actions in the olfactory bulb is notable. It may be recalled that the synaptic potentials in motoneurons are relatively brief; long-lasting responses, as in the case of the recurrent IPSP, are due to repetitive firing of inter-neurons. But in the olfactory bulb, a single volley gives rise to prolonged synaptic actions, as witness the prolonged excitatory period in the mitral tuft (M in Fig. 29) or the IPSP caused by the granule cell input to the mitral dendrites (MC). These per-sistent actions are not associated with impulse firing throughout; they must be due, rather, to prolonged release of transmitter, se-questration of transmitter, or persistent postsynaptic responses to the initial transmitter released. Similar, long-lasting synaptic ac-tions occur in other parts of the brain, as we shall see.

DENDRODENDRITIC RECURRENT INHIBITION The dendrodendritic pathway between mitral and granule cells is one of the most closely investigated pathways in the brain, and it is worth con-sidering the sequence in more detail; see Fig. 30. In (a) is shown the sequence of mitral depolarization (1), granule spine EPSP (2), and mitral IPSP (3). Note that the sequence is such that the IPSP in the mitral dendrite follows the initial brief impulse; it thus occurs in the repolarizing aftermath of the impulse. This se-quence ensures that the impulse and the IPSP do not clash or in some way "short-circuit" each other; the IPSP always follows the impulse, and the repolarization, in fact, sums with the IPSP action.

The spatial aspects of the inhibition are illustrated in Fig. 30(b). Here, it can be seen that the granule EPSP not only provides for self-inhibition of an active mitral cell, but also lateral inhibition of neighboring mitral cells. This is by virtue of the spread of the EPSP to neighboring spines of the same granule tree, and of the numerous connections a single granule cell makes with different mirtal cells. By this means the granule cells mediate both feedfor-ward and feedback inhibitory control of the mitral cells. The spread of activity within the granule tree will be considered fur-ther (below) in relation to the electrotonic properties of the dendrites.

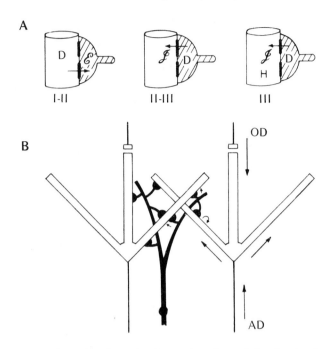

FIG. 30. (A) Postulated mechanisms of action of the dendrodendritic synaptic pathway between mitral (open) and granule (shaded) cells, during successive time periods I, II, and III following an antidromic volley. D, depolarization; H, hyperpolarization; E, excitation; I, inhibition. (B) Diagram of the pathways for self and lateral inhibition through dendrodendritic connections. OD, orthodromic (normal) activation; AD, antidromic activation. (From Rall and Shepherd, 1968.)

The reciprocal dendrodendritic synapses can be seen as providing for recurrent inhibition by a pathway different from that for Renshaw inhibition. The essence of the difference is that the dendrodendritic pathway is local in character; the inhibitory actions of a given mitral cell are confined to the cell itself and to neighboring mitral cells. Even the additional Renshaw-like pathway through the mitral axon collaterals has a local character, in being directed to individual granule spines (see Fig. 28). Recurrent inhibition in the spinal cord, on the other hand, entails the activation of an impulse in the Renshaw interneuron that invades

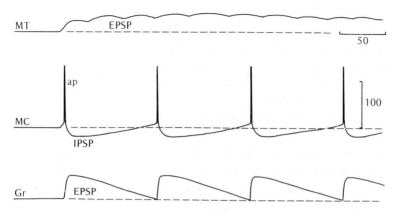

FIG. 31. Postulated mechanism whereby the dendrodendritic pathway may provide for rhythmic activity in the olfactory bulb. Postulated intracellular potentials are shown for the mitral dendritic tuft in the glomerulus (MT), mitral cell body (MC), and granule cell (Gr). See text.

all the axonal branches; it, therefore, has a wide field of action. Because the granule cell has different input-output relations in different parts of the dendritic tree, there is a fractionation of these relations.

RHYTHMIC ACTIVITY   The sequence of dendrodendritic interactions described above is of additional interest in that it provides the basis for the generation of rhythmic activity in the neuronal populations of the bulb. The mechanism will be briefly described (Fig. 31).

As illustrated in Fig. 31, the sequence begins with a long-lasting EPSP in the mitral dendritic tufts (MT) in the glomeruli, due to the olfactory nerve input or the intrinsic activity at the glomerular level. The first mitral impulse generated by the EPSP (MC) synchronously activates all the granule cells with which that mitral cell has synaptic connections (GR). These deliver feedback inhibition of the activated mitral cell, and feedforward inhibition of neighboring inactive mitral cells, in the way already described relative to Fig. 30. As the mitral IPSP subsides, a point is reached at which the EPSP is again at threshold; an impulse is again initiated, and the cycle repeats itself. Through the extensive interconnections between the mitral and granule cells, a steady input in the glomeruli is converted into a rhythmic impulse output in the mitral cell population, locked to a rhythmic activation of the granule cell population.

The activity in these populations generates electric current, which flows through the cells according to the electrotonic properties to be described below. The current flows from the individual neurons summate in the extracellular spaces in and around the olfactory bulb and give rise, thereby, to summed extracellular potentials, which are recorded by an electroencephalograph (EEG). Such rhythmic EEG potentials are a prominent characteristic of the olfactory bulb in the resting state as well as during olfactory-induced activity. Similar mechanisms, although not involving dendrodendritic synaptic connections, have been proposed for the generation of EEG waves in the prepyriform cortex (Chapter 10), thalamus (Chapter 9), and cerebral cortex (Chapter 12); they will be described further in those chapters.

DENDRITIC PROPERTIES

In olfactory bulb neurons, as in the motoneuron, an understanding of synaptic integration must be grounded on an understanding of the electrotonic properties of the dendrites. Because of the prevalence of dendrodendritic synapses, the bulb is a prime model for the study of dendritic properties in relation to dendrodendritic interactions. We will see that these interactions include most of the types found in other brain regions where dendrodendritic synapses occur.

MITRAL CELL   In studying the mitral cell, it is soon apparent that its dendritic tree is not one homogeneous entity, as in the case of the motoneuron. Rather, the mitral dendritic tree is divided into several distinct anatomical entities, and each entity has its own distinct function. The glomerular dendritic tuft is primarily concerned with reception and processing of the olfactory input; it is analogous in this respect to the entire dendritic tree of a thalamic relay neuron (Chapter 9). The primary dendritic shaft, on the other hand, has as its main function the transfer of information from the glomerular tuft to the cell body; it is analogous in this respect to a retinal bipolar cell (Chapter 7). The secondary dendritic branches, finally, are exclusively concerned with interactions with the granule cells.

These divisions are so distinct that it seems as if the mitral cell is not one but three cells, with transfer between them taking place through intraneuronal continuity rather than interneuronal synapses. This means that we must assess dendritic properties in relation to the different functions of each of these entities.

Let us begin with the *glomerular tuft*. We have seen that it forms a small dendritic tree within an olfactory glomerulus. No biophysical model has yet been constructed for the electrotonic properties of this tree, but a few points will indicate in what direction such an effort might lead. The same steps may be taken that were taken for the motoneuron (Fig. 22; see also Figs. 14 and 15). We begin with the initial trunks of the tuft, which have relatively small diameters of 1-3 μm. Let us assume that each dendritic trunk divides in such a way as to conform to the $3/2$ power constraint on the diameter, as discussed in Chapter 4. Each trunk will thus give rise to a small equivalent cylinder, which taken together will form an equivalent cylinder for the entire tuft. Then, assuming a range of values for electrical parameters that is typical of neurons, we can obtain an estimate from the graph of Fig. 13 of a characteristic length of 150–600 μm for the case of 1-μm diameter trunks and 300-1000 μm for 3-μm diameter trunks.

The significance of these estimates is seen when they are compared with the actual extent of the tuft, which is some 150-200 μm. The estimates are thus considerably higher than the actual extents of the tufts. The electrotonic length ($L = x/\lambda$) of an equivalent cylinder for the tuft might, therefore, be estimated at less than 1, and possibly less than 0.5. Note that this is even shorter than the electrotonic length of the equivalent cylinder for the motoneuron dendritic tree. Thus, the smaller branches of the tuft are counterbalanced by their shorter lengths, an expression of the scaling principle previously mentioned in Chapter 5.

Because of the short electrotonic length of the tuft, current flow through the tuft must be relatively effective by passive means alone. The distribution of electrotonic potential may be estimated by referring to the graph of Fig. 13, or Fig. 22, and noting thereon the decrement over a length of $L = 0.5$. It appears that a single synaptic response will spread locally within its branch and neighboring branches to activate the local dendrodendritic pathways through the PG neurons. It also appears that relatively few afferent inputs will be needed to produce a summation of the EPSP's that spread through the tuft to the primary dendritic shaft; the decrement through the tuft, however, must be an important limiting factor in determining the threshold for detection of olfactory signals by the brain. Connections from PG axons and centrifugal axons to the larger glomerular dendritic trunks will have the effect of modulating this afferent transfer through the tuft.

Many of these aspects of organization of the mitral cell tuft are similar to those of the dendritic tree of a thalamic relay neuron (Chapter 9). The comparison between the two is of interest in light of the fact that the olfactory pathway is the only major sensory system that does not pass through a thalamic relay. Perhaps the glomerular layer performs part of that role in the olfactory pathway.

Next to be considered is the *primary dendrite*. This is a single unbranched process, and it is, therefore, easy to make a model for it. From the graph of Fig. 13, it can be seen that likely estimates for the characteristic length of a typical primary dendrite of 6 μm diameter fall in the range of 300-1500 μm. Since a primary dendrite has a length of some 400 μm, it appears that the electrotonic length of this example is at the most 1, and perhaps even less than 0.5. Electrotonic spread should, therefore, be relatively effective through such a process. Apparently, the relatively large diameter of the primary dendrite helps overcome the considerable distance required for the transmission of signals across the EPL.

Since all primary dendrites have roughly the same lengths, thicker dendrites will necessarily have shorter electrotonic lengths; for example, a dendrite of 12 μm diameter would probably have an L length of less than 0.5. In such a case, one might expect electrotonic spread alone to be sufficient for signal transmission. On the other hand, a thin dendrite of say 2 μm diameter would have a length L of more than 1; in such a case, significant attentuation of a signal spreading by passive means alone would be expected to occur. Biophysical models relevant to these cases have been constructed and explored (Rall and Shepherd, 1968). It has been suggested that mitral dendrites may vary in their properties, the larger ones having passive, the finer ones active, membranes. There may well also be species differences; we have noted that myelin has been observed wrapped around the distal primary dendrites in primates, but not in lower animals. Possibly the myelin is associated with active properties of the dendrites; alternatively, it may serve to enhance passive spread. More work is needed.

Finally to be considered are the *secondary dendrites*. The key question with regard to this compartment of the mitral dendritic tree has been the extent to which an impulse at the cell body would invade the secondary dendrites and activate the mitral-to-granule synapses. In the course of investigating the extracellular potential fields in the olfactory bulb, a biophysical model was developed for the secondary dendrites that is relevant to just this problem. This investigation provided the basis for the postulation of the existence of dendrodendritic synaptic interactions in the bulb (Rall et al., 1966; Rall and Shepherd, 1968).

The steps for modeling the secondary dendrites follow those already outlined. By these steps, an equivalent cylinder for the tree of secondary dendrites is obtained; it is illustrated in Fig. 32. Individual secondary dendrites are 2-6 μm in diameter and 400-600 μm in length. Their electrotonic lengths, as well as the values for the equivalent cylinder for the entire tree, have been estimated to lie in the range of 0.5 to 1.

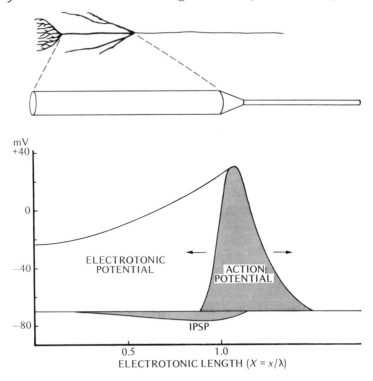

FIG. 32. Electrotonic model of the mitral cell, illustrating the spread of potentials in secondary dendrites.

In the investigation of the properties of secondary dendrites, a model for the action potential was also developed, so that it was possible to simulate an experiment in which an impulse propagates into the cell body and spreads into the dendrites (Rall and Shepherd, 1968). Computational experiments were carried out in which different assumptions were made about the electrotonic lengths of the dendrites and their active and passive properties. The use of biophysical models to perform such experiments that simulate situations often inaccessible in the biological preparation itself is a powerful approach that has been stressed by Rall (1964).

The main result obtained from the model was that impulse spread into the secondary dendrites is very effective by passive means alone. This is illustrated schematically in the diagram of Fig. 32. Passive spread is, in fact, so effective that it was difficult in the computations to distinguish a passively spreading impulse from an actively propagating one, in some cases; this was also true for the primary dendrites. Although this does not answer the question of whether the secondary dendrites have active properties, it does establish the point that an impulse spreading by passive electrotonus retains much of its amplitude and wave form and, therefore, can activate the mitral-to-granule synapses in much the same way a propagating impulse would. Thus, the IPSP produced by feedback from the granule cells would be expected to be distributed throughout the dendritic tree, as indicated in Fig. 32.

In most principal neurons, the dendritic tree receives and integrates both excitation and inhibitory synaptic actions. The mitral secondary dendrites stand out in contrast; they are specialized to receive only inhibition, that inhibition coming from only one type of intrinsic neuron, the granule cell.

GRANULE CELL   Since the granule cell lacks an axon, study of its dendritic properties is obviously crucial to an understanding of its input-output functions.

In the investigation of the recurrent dendrodendritic pathway by Rall and Shepherd (1968), a model for the granule cell was developed and explored. The steps in its construction are summarized in Fig. 33. The branching tree within the EPL is represented by an equivalent cylinder; branch diameters of 0.2-0.8 μm were assumed, an electrotonic length of about 0.4 was estimated. The shaft diameter of the granule cell is of the order of 1 μm; for an average shaft length of 600 μm, an electrotonic length of 1.7 was estimated for the model of the combined tree and shaft. This is significantly longer than the estimates for the different parts of the mitral dendritic tree, but it is similar to the estimates for the motoneuron and the apical dendrites of cortical pyramidal cells (see Chapter 12).

We consider the case of synaptic depolarization of the granule

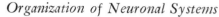

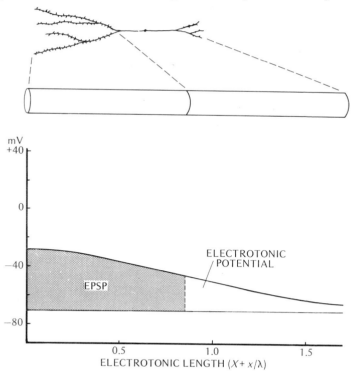

FIG. 33. Electrotonic model of the granule cell, illustrating the spread of potentials for case of EPSP due to a dendrodendritic input to spines in the external plexiform layer.

spines in the EPL, as shown in Fig. 33. The distribution of intracellular potential in the model during the rising phase of the EPSP is shown diagrammatically. The importance of this model in the study of Rall and Shepherd was its demonstration that this synaptic depolarization gives rise to the period III extracellular potentials that are recorded in physiological experiments (Phillips, Powell, and Shepherd, 1963). When the mitral cell model and the granule cell model were joined in sequence, it could be postulated that the EPSP in the spines is due to a dendrodendritic input from the mitral secondary dendrites. As described in the previous section, the localization and timing indicated that the spine EPSP activates inhibitory synapses onto the same secondary dendrites, to produce the long-lasting IPSP recorded in the physiological experiments (Phillips et al., 1963). It was this combined anatomical, physiological, and biophysical study of dendritic properties of the mitral and granule cells that has led to our present understanding of the dendrodendritic synaptic interactions between these cells.

What can be said about the properties of the granule *dendritic spines* within the EPL, spines that have both synaptic input and output? With regard to recurrent self-inhibition of mitral cells, the pathway is entirely local; the granule spines that are activated are the same ones that feed back inhibition onto the mitral dendrites that excited them. With regard to lateral inhibition of mitral cells, we have previously noted that this is mediated by the spread of depolarization from spine to spine within the granule dendritic tree. The distances between neighboring spines are relatively short, some 5-10 μm. Thus, although the stems, or necks, of the spines may be narrow (i.e., 0.2 μm) and the intervening branches thin (i.e., 0.2-0.5 μm), any reasonable assumptions about the electrical properties of the membranes lead to the conclusion that electrotonic spread from spine to spine would be considerable over the short distances involved. It appears therefore that lateral inhibition can be mediated by passive spread alone through the dendritic tree in the EPL. The inhibition is spatially graded according to the electrotonic decrement of the potentials in the tree.

As in the case of the mitral cell dendrites, these considerations do not rule out the possibility of active properties of the granule dendritic membrane, in addition to electrotonic properties. Might the spines themselves have active properties, so that spread into the branches is more effective? And might the branches have active properties, for which there is evidence in cerebellar Purkinje cells (Chapter 8) and hippocampal pyramidal cells (Chapter 11)? The biophysical model for the granule cell does not rule out active spine properties, but it indicates that active properties must be limited and not lead to propagation from the branches in the

EPL into the main dendritic shaft. These questions have been discussed at greater length by Rall and Shepherd (1968) and Shepherd (1972a).

Further study of the properties of the granule cell spines is obviously needed; by virtue of their various strategic locations and their accessibility to experiment, they are excellent subjects for the general study of dendritic spines in the brain. The electrotonic relationship between the spine head and the dendritic branch, through the spine stem, has recently been the subject of analysis by Rall and Rinzel (1971). These points will be discussed further in relation to the dendritic spines of Purkinje cells (Chapter 8), thalamic relay neurons (Chapter 9), and pyramidal cells in the hippocampus (Chapter 11) and neocortex (Chapter 12). In some of these cases the spines have both input and output synaptic connections, whereas, in others, the spines have only input connections. The granule cell is remarkable in having both types. This is an excellent illustration of the principle that, like any other part of a neuron, the functional properties of a spine are determined by the integrative context of the synaptic pathways of which it is a part.

Finally, what may be said with regard to other inputs to the granule cell, particularly the centrifugal inputs from the AON and AC? These inputs make synapses on the shaft and deep processes, and the question arises of how the EPSP's from these inputs spread to activate the output synapses in the EPL. For inputs to the deep processes, perhaps 1 to 2 characteristic lengths (λ) away from the EPL, it would appear that the effect on the synapses in the EPL would be limited if spread through the shaft were by passive means alone. Certainly the amount of activation must be much less than with direct activation of the spines by the reciprocal synapses. The centrifugal inputs, however, have been shown to be capable of giving rise to substantial extracellular field potentials, indicating that spread through the granule shafts is, in fact, effective. Whether this is by virtue of very large EPSP's, or whether spread is assisted by active properties, is not known. Similar questions will arise when the properties of the apical dendrites of pyramidal cells in the hippocampus (Chapter 11) and neocortex (Chapter 12) are discussed.

PERIGLOMERULAR CELL The PG cell has already been discussed as an example of the short-axon cell, a type we will encounter in many parts of the brain. Being small, the PG cell has been relatively inaccessible to experimental studies, nor is there yet a biophysical model as in the cases of the mitral and granule cells. For a first approximation to its over-all electrotonic properties, the dendritic tuft may be regarded as a smaller version of the mitral dendritic tuft described above. Taking into account both the smaller diameters and the shorter lengths of the branches, it seems reasonable to conclude that an equivalent cylinder for the PG cell tuft would be similar to that for the mitral tuft. We assume an electrotonic length (L) of 0.5 to 1 for the equivalent cylinder, as is the case for the mitral tuft. This is, again, an expression of the scaling principle for dendritic trees of different size, mentioned above and in Chapter 5.

As a consequence of this relatively short electrotonic length, spread through the PG cell dendritic tree must be relatively effective. Thus, when many olfactory axons are active and bring their inputs simultaneously to the glomeruli, the resulting EPSP's in the PG cell dendrites not only activate the local inhibitory dendrodendritic synapses onto the mitral cell dendrites, but they also summate and spread to the axon hillock to initiate impulses there. At low levels of olfactory activity, however, when the dew is on the rose, as it were, the synaptic potentials must tend to be restricted to their local dendrodendritic outputs. Under some conditions, the synaptic potentials that give rise to an impulse may facilitate the spread of that impulse back into the dendrites, leading to an enhancement of the dendrodendritic inhibitory output to the mitral cells (Shepherd, 1971a). Such a property should not be a surprise, for we have seen that it is such a spread of the impulse back into the mitral secondary dendrites that activates the dendrodendritic excitatory synapses from these dendrites onto the granule cells.

These aspects of dendritic function in PG cells are similar in many ways to the properties that have been proposed for the short-axon cells of the thalamus; this will be discussed further in Chapter 9.

In discussing the scaling principle relative to motoneurons of different sizes, it was noted that smaller neurons tend to have higher input resistances and larger synaptic potentials; as a consequence, smaller neurons tend to be excitable and have higher rates of spontaneous activity. This is consistent with the activity of olfactory bulb neurons, for the PG cells fire impulses at spontaneous rates of 5-10 sec, in contrast to mitral cells, which fire slowly or not at all. The olfactory input therefore impinges on a glomerular substrate that is under continuous inhibitory bombardment due to the spontaneous activity of the PG cells. This inhibition might be directed to control of the excitability of the mitral cells, or it might be more involved in complex patterns of disinhibitory interactions between the PG cells themselves. An understanding of synaptic dynamics in the olfactory glomerulus, as in other parts of the brain, must take into account this spontaneous activity, which forms the background for the processing of input information.

7

RETINA

The retina is a thin sheet of nervous tissue at the back of the eye, wherein takes place the reception of visual stimuli and the initial stages of processing of information in the visual pathway. The retina differs from the olfactory bulb and from other sensory centers in that the receptors are a part of the neuronal region itself. There is also the obvious difference in geometry between the sheet-like retina and the bulb-like bulb. The flattened conformation of the retina is dictated by the distribution of visual receptors, which in turn is determined by the optics of the eye. These overt features of the retina are striking, but their functional significance can only be assessed in light of an understanding of the internal synaptic organization.

In terms of its main cell types and layers, the retina retains a remarkably similar structure throughout the vertebrate series. It varies considerably, however, in the amount of processing of the visual input carried out within it. On this basis, one distinguishes between simple retinas, in which there is relatively little processing, and complex retinas, in which there is a great deal of processing. This division is notable in not following an evolutionary progression; frogs and rabbits, for example, have complex retinas, whereas primates, including man, have simple retinas (Dowling, 1968; Michael, 1969). The difference reflects the extent to which various types of processing are postponed to higher levels in the central visual pathways.

OUTER SEGMENTS
OUTER NUCLEAR LAYER
OUTER PLEXIFORM LAYER
INNER NUCLEAR LAYER
INNER PLEXIFORM LAYER
GANGLION CELL LAYER
NERVE LAYER

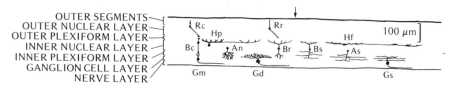

FIG. 34. Neuronal elements of the vertebrate retina. Elements of a simple retina (primate) are shown to left of arrow; some elements of complex retinas are illustrated to the right.

Inputs: cone receptors (Rc) and rod receptors (Rr).

Principal neurons: ganglion cells: midget (Gm), diffuse (Gd), stratified (Gs).

Intrinsic neurons: bipolar cells for cones (Bc), rods (Br), stratified (Bs); Horizontal cells of primate (Hp), of fish (Hf); amacrine cells: narrow field (An), stratified (As).

Retinal layers are shown at left.

NEURONAL ELEMENTS

The over-all similarity in the construction of vertebrate retinas belies the fact that the neuronal elements are differentiated into subtypes, some of which are absent, or show wide variation, in different parts of the same retina, as well as in different species. It is thus impossible to summarize the neuronal elements of the vertebrate retina in one diagram. Our recourse will be to illustrate the main elements of one part of a simple retina, that of the primate, and briefly note comparisons with the elements of complex retinas. The treatment is much oversimplified, and the interested reader should consult Cajal (1911), Polyak (1941), Boycott and Dowling (1969), and Stell (1972) for details.

The neuronal elements of the primate retina are shown in Fig. 34, which is drawn to the same scale as the other diagrams in this book. It emphasizes an important feature of the retina, its extraordinary thinness. The over-all depth is less than 300 μm; apart from the receptor layer, the span of the neuronal elements themselves is less than 200 μm. This is less than the thickness of a single layer in most other regions of the brain. The nerve cells are small and their processes are very limited in the vertical dimension, although in the lateral dimension they may reach extents comparable to those in other brain regions. In the study of synaptic organization, the thinness of the retina cannot be overemphasized. It is the key to understanding the morphology of the neurons and the properties of their dendrites.

INPUTS   The sensory input to the retina is, of course, light. As is well known, the *visual receptors* are of two types: *cones* and *rods*. Cones are specialized for reception of bright light, as in daylight, whereas rods are specialized for reception of dim light, as at night. In most species, the cones also contain the specific pigments that provide for color vision, but there are exceptions to this structure-function relation (see Cohen, 1972).

The receptor cell bodies form a sheet, the outer nuclear layer (ONL) (Fig. 34). Although the receptors are not strictly a part of the neuronal systems of the retina, their properties are crucial to an understanding of retinal organization, as will be pointed out in due course.

The receptors provide the input to the neuronal systems of the retina, analogous to the afferent inputs to the ventral horn and the olfactory bulb. The retina shares with the olfactory bulb the property of having an input of only one sensory modality, although the inputs from cones responding to different wavelengths of light (e.g., color) may be considered submodalities. The input is sent through the process that connects the receptor site in the outer segment through the cell body to the synaptic terminals. In the peripheral part of the retina, this process is vertically oriented and very short (25-50 μm); near the center of vision (fovea), the neuronal elements to which the receptors connect are laterally displaced, so that the processes may run for considerable lengths (up to 0.5 mm). The process is thin (0.5-1.0 μm in rods, 1-2 μm in cones). It is usually referred to as an axon, but, in fact, its fine structure does not permit it to be positively classified as axon or dendrite. This problem of identification will recur with the other elements of the retina.

The receptor terminals are localized within a layer called the outer plexiform layer (OPL) (Fig. 34). In the primate, the cone terminals are large and flattened (5-6 μm across in the central region, up to 15 μm across in the periphery) and are called *pedicles*. Rod terminals on the other hand are small and round (2-4 μm across) and are called *spherules*. This distinction applies generally to other vertebrates, although, in lower forms, the terminals may be more extensive and arborized, and pedicles may arise from rods. The OPL is extremely thin, scarcely more than 10 μm in the primate, as well as in most other species; the terminals of the receptors are thus extremely restricted in vertical extent, more so than for the input to any other region of the brain in our study.

In addition to this afferent input there is in some species a *central* (centrifugal) input from the brain. These fibers run in the optic nerve and terminate in the deeper layers of the retina (Cowan, 1970). These fibers are present in birds, although they are not numerous; in most vertebrate species they are absent or rare. The retina is thus almost exclusively concerned with the processing of its single afferent input; it is unusual among neuronal centers in this respect.

PRINCIPAL NEURON   The output of the retina to the brain is carried in the axons of *ganglion cells*. Their cell bodies lie in a sheet at the inner margin of the retina (see Fig. 34). They vary widely in size and in dendritic arborization. The smallest is the so-called *midget ganglion cell* (Gm in Fig. 34), found near the fovea in the primate. The cell body is 12-15 μm in diameter; each gives off a single dendritic trunk, several microns thick, that extends vertically only some 20-30 μm, ending in several club-shaped terminals scarcely 5 μm across within the inner plexiform layer (IPL). These tiny cells are unique to primates; their cell bodies are among the smallest of principal neurons in the nervous system, and their dendritic "trees" are among the smallest of any neuron. By virtue of their position near the fovea and their restricted extents and synaptic connections (see below), they provide for the greatest visual acuity.

The largest ganglion cells, by contrast, have cell bodies 20-30 μm in diameter. An example is shown in Fig. 34, which has a single trunk (4-8 μm thick) that immediately arborizes over a field of 200-300 μm diameter. Such a cell is termed a giant ganglion cell with *diffuse* dendritic tree (Gd in Fig. 34). Ganglion cells of intermediate size (cell bodies 15-30 μm across) may have diffuse trees of varying extents, or they may have dendrites that branch sparingly at one distinct level; in the latter case, they are called *unistratified* ganglion cells. In non-primate species, ganglion cells may have several distinct levels of branching in their dendritic trees and are referred to as *multistratified* ganglion cells; an example is shown to the right in Fig. 34 (Gs). These cells are characteristic of complex retinas (Maturana, Lettvin, McCulloch, and Pitts, 1960). In some lower forms, the dendritic trees may extend more than 2 mm (Stell and Witkovsky, 1973).

The axons of the ganglion cells pass along the inner surface of the retina and emerge to form the optic nerve. It is especially notable that within the retina the axons give off no recurrent collaterals. The ganglion cells are one of the few types of output neuron in the nervous system that lack such collaterals. The optic nerve projects to the lateral geniculate nucleus of the thalamus, which is described in Chapter 9; there are also projections to the tectum (superior colliculus).

INTRINSIC NEURONS   There are three main types of intrinsic neuron in the retina, each clearly differentiated.

Horizontal cells are the most superficial in location (see Fig. 34). In the primate, their cell bodies are 10-15 μm in diameter; each gives rise to 10 to 15 trunks of varying thickness (1-5 μm). In the central region, the dendritic fields extend only 25 μm in diameter; in the periphery they extend over 100 μm (Boycott and Dowling, 1969; Kolb, 1970; Boycott and Kolb, 1973). The dendrites terminate within the outer plexiform layer. Several of the processes may be especially long and thin. One such process is commonly longer (several hundred microns) and thinner (1 μm diameter) than the rest, and unbranched; by these criteria, it is referred to as an axon. Although these processes are rarely impregnated in Golgi stains throughout their extent, arborizations of terminals, as is illustrated in Fig. 34, are presumed to arise from them. Boycott and his co-workers have provided evidence that the "dendritic" terminals connect only to cone pedicles, whereas the "axonal" terminals connect to rod spherules.

In other vertebrate species, there may be more than one type of horizontal cell; the processes may extend for considerable distances (500 μm or more), and it is often impossible to identify an axon, whether by Golgi impregnation, dye injection, or electron microscopy. In addition to this problem, the horizontal cells have appeared to many observers to be more like glial cells than neurons. In internal fine structure, they have been reported to lack the Nissl substance and the Golgi apparatus typical of neurons and to possess, on the other hand, some of the morphological features of glial cells (e.g., glycogen granules). Some workers have considered them specialized glial "controller" cells. Most modern workers regard them as neurons, on the basis that they have synaptic connections, generate synaptic responses, and are part of a system for processing sensory information. It is doubtful, however, whether any one of these criteria, or all of them together, are adequate to characterize a cell as a neuron; the horizontal cell thus provides a good example of the problem of defining not only the different parts of the neuron but also the neuron itself.

Bipolar cells are oriented vertically and obviously provide the necessary link between the receptor cells and the ganglion cells. They are differentiated into two main types, one of which connects to cone pedicles, the other to rod spherules. A cone bipolar cell (Bc) is illustrated in Fig. 34 (left). It has a small cell body (8-10 μm in diameter), which has earned it the inevitable eponym *midget*. A single peripheral trunk (several microns in diameter) ascends vertically 50 μm or so to terminate in an enlargement some 5-6 μm across; these are much the same dimensions as those of the cone pedicles to which it connects. Depending on the connection to the pedicle, the midget bipolar cells are classified as *invaginating* and *flat;* some of the latter have branches to several cones. We will see the significance of these classifications when we consider the synaptic connections. A single deep process (1-2 μm thick) descends some 50 μm or so from the cell body and ends in several knoblike terminals in the IPL.

A rod bipolar cell (Br) is illustrated in Fig. 34 (right). It also has a small cell body, a somewhat stouter peripheral trunk, and a number of terminals within the OPL, connecting to many rod spherules. The single deep process arborizes sparingly and terminates in knoblike appendages in the IPL.

In different vertebrate species the bipolar cells show many variations, in terms of size, extent of branching, and overlap in connections from the rods and cones (see Witkovsky, 1971; Witkovsky and Stell, 1973). In complex retinas, there may be elaborate arborizations in the IPL, with the branches stratified into several layers and with laterally extending fields of several hundred microns. An example is shown in Fig. 34 (Bs).

It is usual to regard the outer bipolar cell process as a dendrite and the inner process as an axon. Both, however, are similar in fine structure, and there is apparently no axon hillock. One therefore has a problem of identification not unlike that for the receptors and horizontal cells. There seems to be agreement at least that the fine structure of the bipolar cell is characteristic of a neuron.

Amacrine cells are found in the deeper levels of the retina. Their cell bodies are located in the inner nuclear layer (INL), which also contains the bipolar and horizontal cell bodies. There are several subtypes. One has a cell body of 8-10 μm in diameter, and a stout descending trunk, several microns thick, which immediately arborizes into a number of terminals. These are termed *narrow-field diffuse* amacrine cells (An); the field of arborization is typically only 25 μm in diameter. Another type gives off several thin (1-2 μm diameter) dendritic trunks, which branch sparingly and spread over a field of 100 μm or so, although some branches may reach greater lengths. These are termed *wide-field diffuse* cells. A third type has branches confined to one layer in the IPL; these are called, appropriately, *unistratified* cells. There is a resemblance between these branching patterns of amacrine cells and those of ganglion cells. In complex retinas, the amacrine cells show a rich diversity of types, particularly of the *multistratified* variety; an example is shown in Fig. 34 (As).

We have noted the difficulty of identifying an axon in the horizontal and bipolar cells. The amacrine cell has, in fact, been the traditional example in the retina of a cell without an axon. The earliest studies by Cajal and others, using the Golgi method, established this fact, and it has been confirmed repeatedly in modern studies. The analogy with the granule cell of the olfactory bulb in this respect has long been recognized. There is little doubt that the amacrine cell is a neuron, and the fine structure of its processes is consistent with the designation dendrite.

If we take an overview of this community of neurons, it is noteworthy that not only are the neurons differentiated into types, but the types are further differentiated into subtypes, on the basis of distinctive patterns of branching and connection. The retina is unusual in this respect. In the ventral horn, for example (Chapter 5), the motoneurons vary widely in size but not in their basic branching patterns; even the gamma motoneurons, which are functionally distinct, appear for the most part as smaller editions of the alpha motoneurons. Much the same can be said of mitral and tufted cells in the olfactory bulb (Chapter 6). The Purkinje cells in the cerebellum (Chapter 8) have a particularly monotonous appearance. In general it appears that in most local regions of the brain the neurons of a given type differ mainly in terms of size. In the retina, some of the subtypes are related to the rod and cone systems; this is clearly seen in the horizontal and bipolar cell populations, as well as in certain ganglion cells. Some of the subtypes of amacrine and ganglion cells, however, are not obviously related to these systems; they appear to be related, instead, or in addition, to the processing of other aspects of the visual input.

The retinal neurons bear little similarity to the neurons we have encountered in the ventral horn and olfactory bulb, and will meet in other parts of the brain, with respect to either their small size or their peculiar shape. Does this mean that the retina is

Table 2

Quantitative estimates of numbers of optic nerve fibers (from retinal ganglion cells) in different vertebrates.

Lamprey	5,000 (100%)*	Pigeon	1,000,000
Goldfish	53,000	Rat	75,000 (22%)*
Mudpuppy			
(*Necturus*)	360 (100%)*	Cat	100,000
Frog	29,000 (47%)*	Macaque	1,200,000
Chick	400,000	Man	1,000,000

From Bruesch and Arey (1942).
* Per cent unmyelinated fibers.

sui generis? That comparisons with the organization of other regions cannot be made? The peculiarities of the retinal neurons are deceiving in this respect, for they become understandable once one has a grasp of synaptic connections and dendritic properties, and of the functions they serve within the spatial confines of the retina.

CELL POPULATIONS   In man, the following estimates have been made: 100,000,000 rod receptors; 5,000,000 cone receptors; 1,000,-000 ganglion cells. These give input-output ratios for the retina of 100:1 for rods and 5:1 for cones. The latter, presumably, reflects the high acuity associated with the cone systems. In the human fovea, in fact, where visual acuity is highest, it is believed that the ratio of cones to ganglion cells is less than 1:1. The convergence ratios vary widely for different input-output subsystems within the retina, as has been discussed by Stell (1972).

The numbers of neuronal elements vary considerably in different species. Estimates of the number of optic nerve fibers in some representative species are given in Table 2. The figures are a direct measure of the numbers of ganglion cells giving rise to the fibers. It may be seen that there is a general tendency toward larger populations and myelination of the fibers, but there are wide differences between closely related species within that overall framework.

The significance of such numbers depends on many factors, and it is not so obvious as it may appear. For example, Bruesch and Arey (1942) set the 1,000,000 optic nerve fibers against the total of all other craniospinal nerves, and concluded "that 38% of all the fibers entering or leaving the central nervous system do so by way of the optic nerve." This was considered to be a quantitative evaluation of the "predominant role of vision in human activities"; Granit (1955) cited this figure to the effect that "we are exceptionally visual animals." But the calculation, for some reason, neglected the 50,000,000 olfactory nerve fibers; if they are included, is it evidence that we are "exceptionally olfactory" as well? And consider further that there are only some 30,000 auditory fibers to carry the input from the ear to the brain. Is this evidence that the auditory input is insignificant by comparison? Although these numbers are important in reconstructing the mechanisms of information processing in the different sensory systems, they must be viewed with great perspicacity in assessing the relative roles of the systems in the lives of the species.

One figure in Table 2 is particularly striking, and that is the very small number of optic nerve fibers (*ergo* ganglion cells) in the mudpuppy (*Necturus*). One might expect that with so few output channels this would be, perforce, a simple retina, but on the contrary it is a complex retina, much more so than the human retina with almost 3000 times as many cells! This reflects the important principle that functional complexity depends on synaptic connections, not on number of neurons.

We may mention, finally, that numbers for intrinsic neuronal elements are hard to come by. It may be estimated that the ratio of bipolar to ganglion cells is of the order of 2:1; of horizontal cells to ganglion cells is 5:1; and of amacrine cells is 10-100:1. These are only very crude estimates, and they vary with the species and complexity of the retina. It is at least clear from these ratios that the populations of intrinsic neurons are relatively large and that a great deal of processing goes on through the connections that they make, particularly at the amacrine cell level. In this respect the retina is similar to the olfactory bulb and different from other regions with relatively low ratios of intrinsic to input and output elements.

SYNAPTIC CONNECTIONS

The study of synaptic connections in the retina has been an extremely active area of research. The identification of terminals has been aided by the sharp differentiation of cell types and the localization of synapses in the outer and inner plexiform layers. The development of techniques for serial reconstructions of electron micrographs (Sjöstrand, 1958; Missotten, 1965; Allen, 1969; Kolb, 1970) and for fine structural analyses of Golgi-stained neurons (Stell, 1964, 1967) has put identification on a sound basis, and has permitted quantitative analyses of some connections to be made. We will again focus our attention on the salient features of the primate retina, with brief mention of other species (for reviews, see Dowling and Boycott, 1966; Stell, 1972).

OUTER PLEXIFORM LAYER The OPL contains the terminals from the receptor cells, bipolar cells, and horizontal cells. A cone pedicle is shown in (A) of Fig. 35 and a rod spherule in (B). Note the size of the cone pedicle; it is among the largest terminal structures in the nervous system [compare the terminals of mossy fibers in the cerebellum (Chapter 8)]. That these large structures should arise from the small receptors illustrates the fact that terminal size is not necessarily related to cell body size.

The receptor terminals have invaginations that envelop the terminal processes of both bipolar and horizontal cells. The patterns of connection between the terminals are characteristic for rods and cones. In primates, the invaginations in the *rod spherule* contain a central process of the dendrite of a rod bipolar cell, flanked by two processes from the "axonal" arborizations of horizontal cells. This is shown in Fig. 35(A). The three processes within an invagination have been termed a *triad* (Missotten, 1965; Stell, 1967), not to be confused with our use of that term to apply to general principles of organization (see below). *Cone pedicles*

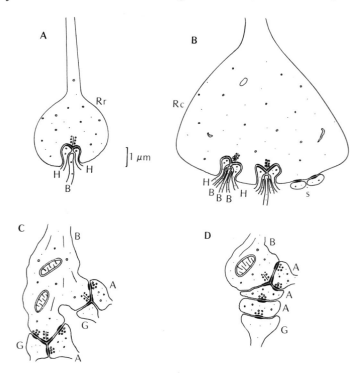

FIG. 35. Synaptic connections in the primate retina. (A) Connections in the outer plexiform layer, of a rod receptor (Rr) into bipolar (B) and horizontal (H) terminals. (B) Connections of a cone receptor (Rc); note also superficial (s) contacts. (After Kolb, 1970; Cohen, 1972.) (C) Connections in the inner plexiform layer, between bipolar terminals and amacrine (A) and ganglion (G) cell dendritic terminals. (After Dowling and Boycott, 1966; Allen, 1969.) (D) For comparison with (C), connections in the inner plexiform layer of a complex retina (frog). (After Dowling, 1968.)

have invaginations in which a dendrite from an invaginating midget bipolar cell is flanked by two processes from the "dendritic" arborizations of horizontal cells [Fig. 35(B)]. Clustered around the bipolar cell dendrite are terminals from several other bipolar cells of the flat midget and flat variety. This may be said to form a *synaptic complex*, in the sense in which that term is used elsewhere in the brain (see Chapter 8). The careful reconstructions of Kolb (1970) have shown that, in the central retina, a single invaginating midget bipolar cell connects to a single cone pedicle by means of 10 to 25 triads. The multiple and exclusive nature of this synaptic relationship is remarkable, although not unique; we have previously noted the multiple contacts of Ia terminals onto motoneuron dendrites (Chapter 5). Flat midget bipolar cells connect to approximately 6 cones, whereas rod bipolar cells connect to 10 to 50 rod spherules.

Let us now consider the synaptic junctions more closely. Within the receptor terminal, and opposite the point at which a bipolar and horizontal cell triad of processes meet, is a so-called *synaptic lamella* (Ladman, 1958) or *ribbon*, surrounded by a cluster of vesicles. This is indicated in Fig. 35(A,B). There is a moderate amount of densification of the pre- and postsynaptic membranes in the vicinity of the ribbon. The horizontal cell processes of the triad are in direct apposition to the terminal membrane, but the central bipolar cell process is some 800-1000 Å distant, e.g., roughly 1 μm (Dowling and Boycott, 1966). It is not known, in fact, whether synaptic transmission occurs directly from the terminal to the bipolar cell process, or indirectly through the horizontal cell processes. These junctions are, therefore, quite distinctive; they are specialized synapses, in the sense we have previously discussed (Chapter 2). It is clearly at this site that information transfer takes place, and the synaptic ribbon with its vesicles is taken as presumptive evidence that the transfer takes place at least in part by chemical means.

Many other types of synapses are found in the OPL. There are contacts by the flat bipolar cells on the outer faces of the terminals; these are termed *superficial* contacts. In lower vertebrates, simple ("conventional") synapses are observed from horizontal cells to bipolar cells, and between horizontal cells. Gap junctions are present between horizontal cells in some retinas, particularly in fish (Yamada and Ishikawa, 1966); this is significant relative to physiological evidence for electrotonic coupling between horizontal cells in those retinas (see below). Finally, there is evidence in some retinas for contacts between receptors and from horizontal cells back onto receptors (see Stell, 1972, for review). The retina is special in this respect; synapses back onto the input elements are not found in the other regions of the brain in our study.

INNER PLEXIFORM LAYER   The second level for synaptic connections is the IPL, which contains the terminals of bipolar, amacrine, and ganglion cells. Ribbon synapses are again encountered, here located within the bipolar cell terminal (Dowling and Boycott, 1966; Dowling, 1968). It is noteworthy that ribbon synapses are found within two cell types that are disparate in morphology and have distinct yet very closely linked functions. Note also a large terminal arising from a small neuron; a bipolar cell terminal may be up to 8 μm in diameter. The presynaptic ribbon in a bipolar terminal is situated opposite the point where two postsynaptic processes meet; hence, the junction is called a *dyad*.

The identity of the postsynaptic processes varies according to the complexity of organization. In simple retinas (e.g., primate), one of these terminals is from an amacrine cell, the other is from a ganglion cell. Typically, the amacrine cell has, in addition, a synapse back onto the bipolar terminal. The term *reciprocal synapse* was first used to describe this arrangement (Dowling and Boycott, 1966), and it has been generally accepted in describing the similar arrangements of side-by-side synapses in other parts of the brain (see Chapters 2, 6 and 9). Amacrine cells also have synapses onto ganglion cell dendrites. These patterns of connections are illustrated in Fig. 35(C). By serial reconstructions, Allen (1969) has shown that a single elongated terminal of a bipolar cell has as many as 23 synaptic ribbons, at which sites connections are frequently made to more than two other processes. Stell (1972) has noted that these complex clusters of processes resemble, in certain respects, the synaptic glomeruli of other parts of the nervous system (see Chapters 8 and 9).

In complex retinas (e.g., frog), all the processes opposite a ribbon may belong to amacrine cells, and there are numerous amacrine-amacrine synapses of the simple ("conventional") type. It has been inferred that, in these retinas, the bipolar cells connect to the ganglion cells not directly but through several amacrine cells, as illustrated in Fig. 35(D). The term *serial synapse* was first used by Kidd (1962) to describe these sequences. The finding of amacrine-amacrine synapses is very important evidence that an intrinsic neuron can have synapses onto other intrinsic neurons of the same type. The complexity of the amacrine-amacrine sequences appears to be directly related to the complexity of information processing (Dowling, 1968). In the cerebellum (Chapter 8) and the cerebral cortex (Chapters 10 through 12), we will see that principal neurons can have synapses onto other principal neurons of the same type.

We may conclude that the synaptic connections in the retina show a great diversity of types. Some can be characterized as specialized terminals and synapses, others as simple terminals and synapses. Since many of the intrinsic neurons lack a clearly defined axon, the terminology for the types of synapses, in terms of the processes that take part in the synapses, is not settled. One extreme position would be that all the synapses can be regarded as dendrodendritic, or at least equivalent to that type of connection. At the other extreme, some of the synapses would be regarded as axodendritic, as, for example, from the bipolar terminal onto the ganglion cell and amacrine cell dendrites [Fig. 35(C)]. But this immediately requires that one label as dendroaxonic the reciprocal synapses from the amacrine cell to the bipolar terminal. It also requires that one label as axon a process that does not generate an action potential (see later). Thus, under any set of definitions, one obtains types of synaptic orientation that violate the functional canons of the traditional neuron doctrine, as outlined previously in Chapter 1. The retina, therefore, furnishes excellent proof of the need to replace those canons with more flexible concepts of synaptic organization.

BASIC CIRCUIT

The knowledge of synaptic connections between the clearly defined neuronal types of the retina leads directly to a summary of organization in a basic circuit diagram, as in Fig. 36(A).

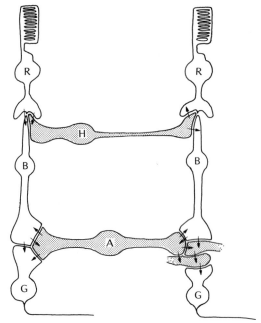

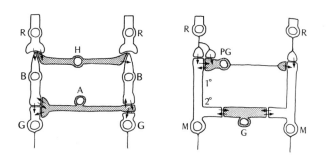

FIG. 36. Basic circuit diagram for the vertebrate retina. Abbreviations as in Fig. 35. (Modified from Dowling, 1968.) *Below,* comparison between the synaptic organization of the retina and olfactory bulb.

To summarize briefly: in the OPL, the receptors connect to both bipolar and horizontal cells through the ribbon synapses. A horizontal cell may, in turn, have synapses with bipolar cells, with other horizontal cells, and also back onto receptor terminals. In the IPL, the bipolar terminals connect to both ganglion cells and amacrine cells through their ribbon synapses. The amacrine cells have reciprocal synapses back onto the bipolar terminals and other synapses onto ganglion cells. This pattern is characteristic for simple retinas, as shown in Fig. 36(A). To the right in the figure are shown the serial synapses through amacrine cell processes that are characteristic of complex retinas.

It can be seen that the main framework of the retina is one of vertical and horizontal pathways. There are vertical pathways from receptors through bipolar cells to ganglion cells, and horizontal pathways between the vertical paths. The horizontal connections are separated rigidly into two levels, or tiers. At the first level, the OPL, the horizontal cells are related to processing of the receptor input to the bipolar cells. At the second level, the IPL, the amacrine cells are appropriately situated to control the output from the ganglion cells.

In terms of this framework, the retina appears to be organized on principles similar to those of the olfactory bulb. A diagram for the bulb is included for comparison [Fig. 36(B)]. The framework of vertical and horizontal connections and the two-tier separation of the horizontal connections are obviously similar. The horizontal cells are analogous to the PG cells in being at the level of input processing, whereas the amacrine cells are analogous to the granule cells in being at the level of output control. Note that the amacrine cells and granule cells both lack axons and both take part in reciprocal synaptic connections with the vertical pathway at the output level.

These similarities indicate that the over-all outline of synaptic organization in the retina is not unique. It is useful to view the retina in this light, for it is one of the steps that is necessary in the development of common principles that apply throughout the brain.

Let us now consider the organization of the retina more closely, using the diagram of Fig. 36 as our guide. In discussing the olfactory bulb, we introduced the principle that the input elements to a region are related in some way to both the vertically and the horizontally conducting elements in that region; the three types of elements form, it was suggested, a synaptic triad. In the retina, the ribbon synapses of the receptor terminals provide a simultaneous input to both the horizontal cells and the vertically conducting bipolar cells. This type of synaptic arrangement is similar to that in the cerebellum (Chapter 8) and thalamus (Chapter 9), and appears to provide for a more synchronous and possibly more stereotyped transfer to the vertical and horizontal pathways than is the case in the olfactory bulb (see Chapter 6). This might be essential for the transfer of spatio-temporal patterns and sequences in the retina.

The position of the bipolar cell as the vertical conducting element from input to output levels is clearly shown in the diagram of Fig. 36. In this function, the bipolar cell is clearly analogous to the primary dendrite of the mitral cell in the olfactory bulb. Note that we have compared, in this way, the function of an entire neuron (the bipolar cell) with the function of only part of another (the mitral cell). Consider, as an alternative, that the anatomically bounded cell is the actual unit of function. Then the entire mitral cell must be comparable to the entire bipolar cell, since both are the first-order neurons in their respective pathways. This comparison is obviously of limited value. In the analysis of the functional significance of synaptic organization, therefore, functional entities must be identified regardless of anatomical boundaries, a principle that has already been introduced in discussing the neurons of the olfactory bulb (Chapter 6).

If we next consider the connections of the bipolar cells in the IPL, it can be seen that the relation of the ribbon synapses to the terminals of ganglion cells and amacrine cells is similar to the triadic pattern in the OPL. In complex retinas, this input is largely to the amacrine cells alone, i.e., the intrinsic elements (see Fig. 36). The break, as it were, in the vertical pathway at this level, provides for more degrees of freedom than is possible in the case of direct continuity from primary dendrite to mitral cell body in the olfactory bulb [see Fig. 36(B)]. These greater degrees of

freedom are effected by increased synaptic interconections involving the intrinsic elements, as is shown by the high degree of
correlation between the functional complexity of the r a and
the number of amacrine-amacrine synapses (Dowling, 196 and

The amacrine cells exert their control over the ganglio.
in three ways: through synapses onto the bipolar cell term s
by synapses directly onto the ganglion cells, and by amacr.
amacrine connections. The control mediated by the amacrin
bipolar synapse (which is, in fact, part of the reciprocal pair) i.
of particular interest, since it is presynaptic in position relative to
the ganglion cell. It is, thus, part of the functional repertoire of
the IPL that the amacrine cells can control the ganglion cell out
put through both presynaptic and postsynaptic connections. Th
fact that the ganglion cell is entirely postsynaptic in position (i.e.,
its dendrites have no presynaptic functions) means that, by itself,
it is a relatively simple integrative unit, like the motoneuron, as
compared with a neuron like the mitral cell of the olfactory bulb.
Its true complexity, however, is seen when it is considered as part
of the larger multineuronal functional units it forms with the bipolar and amacrine cells.

The synaptic connections at the two levels in the retina provide many pathways for feedback and feedforward interactions.
Compared to other local regions, the retina is notable in being de-
void of internal feedback loops through axon collaterals from the
output neurons and external feedback loops from other parts of
the brain. It is as if the retina has quite enough to do, thank you,
handling the visual input alone. In the OPL, the connections are
specific in many retinas for rod and cone systems, although this
is not shown in the basic diagram of Fig. 36(A). In the IPL, a
given synaptic connection may be part of either, or both, systems, as well as part of other systems that process more complex
aspects of the visual stimulus. It seems highly likely that many of
these pathways overlap, and that a synapse is multifunctional in
the sense previously discussed in Chapter 6.

The sequence of neurons and layers in the retina is non-
repeating, as is characteristic of cortical structures in the terms

of our disc~~ussion~~ in the preceding chapter. The anatomical sepa-
rateness o~~f~~ ~~t~~he bipolar cell in the vertical pathway means that the
output ~~neu~~ron, the ganglion cell, does not span the layers of the
retina~~l syste~~m; characteristic of output neurons in the olfactory bulb,
cere~~brum~~, and other cortical regions. The larger functional unit
fo~~rmed~~ by the bipolar and ganglion cell together does span the
~~retina~~e, from the input to the output level. In this aspect, as well
~~as in~~ many others, the retina represents an extreme variation on
~~the~~ basic plan of cortical organization.

SYNAPTIC ACTIONS

It is a great advantage in the investigation of the retina that one
can not only activate it with a specific stimulus (light) but activate it discretely, with stimuli closely controlled in spatial extent
and temporal course. No other region of the brain provides these
advantages for the experimenter to the extent that the retina does.
One can, therefore, analyze the retina using light stimuli rather
than electrical shocks, and the results are correspondingly more
closely relevant to the processing of natural stimuli. The basic
types of stimuli used are a spot and a surrounding rim (annulus)
of light.

The disadvantage in the study of the retina is the small size of
the neuronal elements; analysis has had to await the development
of refined techniques for intracellular recording and staining (see
Svaetichin, 1953; Tomita, 1965, 1972; Werblin and Dowling,
1969; Kaneko, 1970). The results that are considered typical for
synaptic actions in the different types of neuron in the retina are
illustrated in Fig. 37. In describing them, we will begin with the
receptors and work inward to the ganglion cells.

Intracellular recordings from a *receptor cell* (R) are shown in
Fig. 37. The salient features are, first, that the resting membrane
potential is very low (10-30 mV); second, the response to a spot
of light is a slow potential transient (not an impulse); and third,
the potential is in the hyperpolarizing direction. Rod receptors
respond in graded fashion proportional to light intensity. Cone

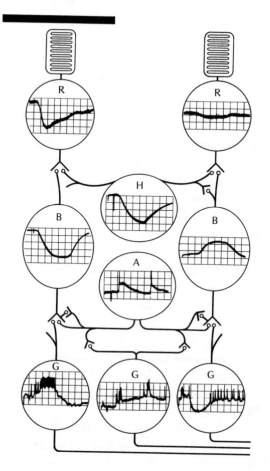

FIG. 37. Synaptic actions in the vertebrate retina, as recorded intracellularly from neurons in *Necturus* (mudpuppy). *Left,* responses recorded at the center of a spot of light (bar above). *Right,* responses in the surround. Voltage calibrations, one scale division equals 1 mV (R); 2 mV (H, B, and G); 5 mV (A). Time calibration, one division equals 200 msec. (From Dowling, 1970, after Werblin and Dowling, 1969.)

receptors are of three types depending on their peak sensitivities to blue, green, and red light.

It is a general characteristic of sensory receptors that they respond to their specific stimulus by a depolarization; the mechanism for this depolarization involves a conductance increase similar to that for an EPSP. The hyperpolarizing responses of receptor cells in the vertebrate retina were therefore astonishing when revealed in the early electrophysiological investigations (see Tomita, 1965, 1972). The explanation appears to lie in the fact that, in the resting state, there is a constant leak of electrical current from the light-sensitive region of the receptor (the outer segment), which keeps the resting membrane potential at its relatively low level. The effect of a sudden light stimulus is to reduce this current and hyperpolarize the cell. There is evidence that the dark current is carried principally by Na ions, and that light acts by reducing the permeability to Na at some membrane site in the outer segment (Penn and Hagins, 1969; Toyoda, Nosaki, and Tomita, 1969).

Horizontal cells (H in Fig. 37) also give slow responses to light stimuli. As in receptor cells, the responses arise from low resting potentials. These were the first intracellular recordings from the retina (Svaetichin, 1953); the responses are termed S (slow) potentials. The responses are of two types: the luminosity type (L), a hyperpolarization of the cell to all wavelengths of light, and the chromaticity type (C) (not indicated in Fig. 37), in which the response may be hyperpolarizing or depolarizing, depending on the wavelength of light.

What is the mechanism for synaptic transmission from the receptors to the horizontal cells? Figure 37 shows that the potential change in the presynaptic terminal of the receptors is a hyperpolarization and that the postsynaptic potential change in the horizontal cell is also a hyperpolarization. This poses a difficult problem, because, as Hodgkin (1971) has pointed out, there are no known exceptions to the rule that transmitters are released by depolarization of the cell membrane. This requires the postulate that the receptors release transmitter continuously in the dark and that light suppresses this release. There is evidence that this is, in fact, the case. In response to light, the horizontal cells undergo a decrease in conductance similar to that of the receptor cells (Toyoda et al., 1969), so that it appears possible that the mechanism of the response of horizontal cells may be similar in principle to that of the receptors, as described above. Recent studies implicate an amino acid (e.g., aspartate) as the transmitter substance from the receptors to the horizontal cells (Dowling and Ripps, 1972).

It should be clear to the reader at this point that the hyperpolarizing responses in receptors and horizontal cells are due to a mechanism that is different from that of the hyperpolarizing IPSP's found in many neurons in other parts of the brain. They are produced by a turning off of a depolarizing conductance channel (i.e., Na), whereas the IPSP is produced by a turning on of a hyperpolarizing conductance channel (i.e., K). That, at least, is the mechanism that has been postulated for the horizontal cells (see Trifonov, 1968), although recently other possibilities have been considered (Nelson, 1973). See Chapter 3 for a discussion of these and other synaptic mechanisms. The postsynaptic hyperpolarizing potential in the horizontal cell is, thus, presumed to be produced by a sudden interruption of synaptic depolarization from the receptors, rather than by an active inhibitory process. The effect is actually similar to that produced by presynaptic inhibition; that is, a suppression of excitatory drive without active inhibition of the cell driven (cf. Fig. 24).

Bipolar cells, like receptors and horizontal cells, also generate only slow potentials in response to a light stimulus (see B, Fig. 37). The latency of onset of the response is similar to that of the horizontal cell, which is consistent with the idea that the receptor terminal, through its ribbon synapse, provides for simultaneous transmission to both cell types. The responses also arise from resting potentials that are relatively low, in the range of 30-40 mV. The responses may be hyperpolarizing (example, left, Fig. 37) or depolarizing (right, Fig. 37). Different bipolar cells show different polarities in their responses to a central spot and a surrounding annulus of light flashed on the retina. The bipolar cell is the first in the sequence of neurons within the retina to show marked center-surround antagonism by opposite polarities of response in relation to spatial aspects of a stimulus.

Amacrine cells respond to light stimuli with graded depolarizing potentials upon which may be superimposed one or two spikes (A, Fig. 37). Amacrine cells are therefore the first in the sequence of neurons within the retina to respond with depolarizations resembling the EPSP's of other neurons in the brain and the first to generate impulse activity. The case of an amacrine cell that responds to both the ON and the OFF of a spot of light with slow potentials and single impulses is shown in the example in Fig. 37. Amacrine cells vary in whether they respond at ON or OFF and in the grading of the amplitudes of the responses. The responses are relatively transient compared with the more sustained responses of the other cell types described earlier.

Ganglion cells respond to light stimuli with various combinations of EPSP's, IPSP's, and impulse discharges, all of which resemble their counterparts in other output neurons of the brain. Resting membrane potentials are relatively low (40-50 mV). Ganglion cells are the first (and only) neurons in the sequence within the retina to show spontaneous impulse activity. Some ganglion cells respond with a steady discharge to a central spot of light, and are inhibited by a surrounding annulus (left and right G, Fig. 37); it has been suggested that these ganglion cells may receive their input primarily by way of direct connections from the bipolar cells, which have similar receptive field organization (see Dowling, 1970). Other ganglion cells show transient responses at both the onset and cessation of a light stimulus (center G, Fig. 37); this is similar to the pattern of responses in amacrine cells, and suggests that such ganglion cells are driven predominantly through amacrine cells.

Figure 37 gives only a slight hint of the range of responses that has been recorded in ganglion cells. In simple retinas, the ganglion cells respond primarily to stationary spots and annuli of light, as was shown in the classical experiments of Kuffler (1953). In complex retinas, the ganglion cells are less responsive to simple and stationary stimuli and more responsive to complicated aspects of the stimuli (Barlow, 1953; Maturana et al., 1960). For example, a ganglion cell may respond only to a moving spot of light, or to movement in one particular direction. Or it may respond, not to a spot of light, but to an edge of light and dark; to evoke a maximal response, the edge may have to be convex or concave.

The nature of these responses suggests that the retina abstracts specific aspects of the stimuli and that ganglion cells are tuned, as it were, to these aspects. Since such responses are never seen in the more peripheral retinal elements (receptors and horizontal and bipolar cells), it means that the tuning must come about through synaptic interactions mediated by the amacrine cells in the IPL. This is consistent with the fact, noted above, that complex retinas are characterized by a well-developed IPL, containing dendritic trees of ganglion and amacrine cells with complicated branching patterns, and a multiplicity of connections between the amacrine cells. Within this neuropil are established the excitatory and inhibitory feedforward and feedback pathways that are the basis for the functional properties of the ganglion cells.

If we take an overview of synaptic actions in the retina, the outstanding features seem to be the slow graded potentials and lack of impulses. It may seem paradoxical that the basic circuit of the retina should have many points in common with other circuits in the brain while the retinal neurons should be so idiosyncratic in their functional properties and synaptic actions. This need not be perplexing, however, if we recognize that information can be processed by different mechanisms to the same end; analog and digital computers provide familiar examples of this truism. It is just so in the nervous system; similar operations are performed by different neuronal structures with different properties, and these similar operations are implied to a certain extent in the interactions as they are embodied in the basic circuits. In these remarks, we touch on important principles of structure-function relations in the neurons of the brain, which will be discussed further in the next section.

DENDRITIC PROPERTIES

It has been shown that the intrinsic neurons of the retina are remarkable not only for their small size but also for the paucity or lack of morphological axons. The study of these neurons can, therefore, be said to involve largely a study of their dendritic properties. It has also been shown that there is a dramatic absence of impulse traffic within the retina. This means that the spread of activity through the retinal neurons is by graded passive current flow, produced by the slow potential responses. We are, by now, well aware that such spread of activity is determined by electrotonic properties. Knowledge of these properties is, therefore, especially important in the study of the retina, as the necessary basis for reconstructing functional organization.

It would be hoped in the light of these remarks that electrotonic models were available for retinal neurons, but this is unfortunately not the case. A model has been developed for responses of visual receptor cells, and it will be described. With regard to the other types of neuron, we will only be able to note some preliminary steps toward the construction of models and to discuss some implications of dendritic structure and function.

RETINAL RECEPTORS   An important problem in vision research is the determination of the least amount of light that can be detected. The behavioral threshold for perception of light is at the level of a few quanta, each quantum activating a separate rod receptor (Hecht, Schlaer, and Pirenne, 1942). The threshold for detection by a single rod is, therefore, at its absolute limit, one photon (quantum) breaking down one molecule of visual pigment in one rod. The mechanism for this transduction process is itself a fascinating one, but our interest is in the question most pertinent to the organization of the retina: How can the response to one photon produce a change in the presynaptic terminal of the receptor that will be of sufficient magnitude to provide for transmission to the rest of the retina and, ultimately, the brain?

In experiments on the nature of the dark current in rod receptors of the rat, Hagins, Penn, and Yoshikami (1970) have asked just this question and have attempted to answer it by constructing an electrotonic model for the receptor cell. Because the receptor cell changes diameter markedly at several levels it cannot be ap-

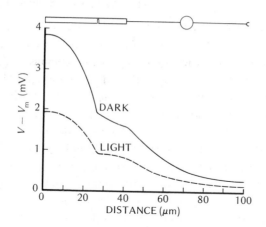

FIG. 38. Electrotonic model of a retinal receptor cell (rat), to illustrate the distribution of the membrane potential at rest (dark) and during a saturating flash of light (light). (From Hagins et al., 1970.)

proximated by a simple cylinder, but rather, as shown in Fig. 38, by a cylinder with segments of different diameters. The lengths and diameters for the different segments are indicated in Fig. 38.

The longest part over which the response must spread is from the inner segment through the cell body to the presynaptic terminal. This part has a length of about 50 μm and a diameter of about 0.4 μm (except for the region of the perikaryon). From our previous diagram (Fig. 13), it can be seen that this would indicate a characteristic length (λ) of 50-200 μm, using the common values of electrical parameters, which means that the electrotonic length (L) of the receptor would be less than 1 and passive spread would be very effective. The experimental evidence, however, is that the specific membrane resistance (R_m) is, in fact, very low, of the order of only 100 Ωcm^2, for which a λ value of only about 25 μm was calculated (cf. Fig. 13). Thus, the rod receptor has a relatively long electrotonic length, approximately $L = 2$. From previous discussion, we realize that this is not too long to prevent passive spread of large potentials, but it severely restricts the spread of a threshold response.

The decrement of passive potential through the receptor was calculated by Hagins et al. for two cases: the case of steady current in the dark and the case of a maximal response to a flash of light. The results are shown in the graph of Fig. 38. The change of potential at the synaptic terminal is from 320 μV (in the dark) to 160 μV (in the light). The numbers of photons were calculated for the flash of light, from which was derived the change in potential at the synaptic terminal per single photon. This was calculated to be about 3.6 μV. It was concluded that this response was sufficiently above noise level (about 0.9 μV) to be reliably detectable at the synapse. From this Hagins et al. concluded that "the photocurrent produced by the rod receptor is large enough to permit single photon detection, even with the short space constant and high membrane conductance found for rat rods in our experiments."

The diagram of Fig. 38 also illustrates the standing voltage gradient through the rod receptor. What is the significance of the dark current and this associated voltage gradient? One possibility suggested by Hagins et al. is that it serves as a DC signal carrier, which is modulated by the response to light. Another is that it may be related to transport of materials within the receptor. That the receptor is very active metabolically, and might require such transport, is indicated by the high density of mitochondria within it. In these respects, the receptor is a good example of the special kinds of properties that are present, or may be present, in the small processes of local regions of neuropil, as discussed in Chapter 3.

BIPOLAR CELLS   There are two main aspects of bipolar function, one local and integrative, the other transmissive. The local role is carried out at both input (OPL) and output (IPL) levels in the retina. At both levels, the bipolar cell branches have synaptic inputs as well as outputs, so that a single branch or terminal functions to some extent as an input-output unit. The degree of autonomy of such units depends on the amount of electrotonic current spread between them. In view of the rather limited branching fields, it would appear that there is fairly effective spread of local synaptic inputs to neighboring branches; a local synaptic input would, therefore, lead to output that was graded in both intensity and in local spatial extent.

The other aspect of function is the role of the bipolar cell in transferring signals vertically from input to output level. We have noted the analogy to the mitral primary dendrite in this regard. The transfer in the bipolar cell is apparently exclusively by passive electrotonic spread. The problem of assessing the amount of transfer is not unlike that in the retinal receptor cell, as described above. There is no electrotonic model available for the bipolar cell, but insofar as it can be considered as a single process extending from OPL to IPL, one can indicate how such a model might be developed. Experimental measurements (Nelson, 1973) suggest that the electrical properties of bipolar cell place it between the middle and lower lines of the graph of Fig. 13. Assuming a diameter of 1 μm, the process would have a characteristic length of 150-300 μm. Since the bipolar cell is approximately 100 μm in vertical extent, its electrotonic length (L) would be considerably less than 1, and it could be concluded that spread would be very effective through it.

At higher levels of input intensity, the transfer from OPL to IPL is quite effective, as evidenced by the large potentials recorded from the cell body. Transfer in the opposite direction, from IPL to OPL, however, is presumably also effective, and this is relevant because of the inputs from the amacrine cells onto the bipolar terminals in the IPL. To what extent can the responses to these inputs spread back through the bipolar cells, to influence synaptic outputs from the bipolar terminals in the OPL? If it is assumed that this kind of "backtalk" must be held to a minimum, then one has the interesting proposition that the dendritic properties of the bipolar cell represent a compromise between minimizing transfer in this direction while maximizing it in the other. There are possibly similar principles underlying the properties of other dendrites within the overlapping multi-functional local circuits in the brain.

HORIZONTAL CELLS   In the early work on S potentials, it was clear that the recording electrode was within a confined space, but whether it was a cell body or an extracellular compartment was difficult to determine. Various explanations were tendered for an extracellular compartment whose resting potentials could be the negative sum of the neighboring cells. The image conjured up by Naka and Rushton (1967) was that of a river, "flowing through that crowded community into which all cells can empty their electrical effluence. The latter is a rather versatile concept, but degrading. One likes to have one's electrode in the council chamber not in the sewer." Fortunately, the electrode has been found to be in the council chamber (Werblin and Dowling, 1969; Kaneko, 1970; Baylor et al., 1971).

We have noted the strategic position of the horizontal cell terminals, which are interposed, to a certain extent, between the receptor terminal and the bipolar cell dendrite [Fig. 35(A,B)]. It is possible that the horizontal cell processes have both presynaptic and postsynaptic positions and function, therefore, as local input-output units. Spread of synaptic potentials within the processes is presumably by passive means alone, graded in intensity and spatial extent, as in the bipolar cell terminals.

The horizontal cells also provide for the transfer of local inputs to other parts of their branching trees by passive electrotonic spread. In the primate horizontal cell, the single long slender "axon" is particularly enigmatic. If one is indeed justified in labeling it an axon, then it stands as a rare example of an axon that does not conduct action potentials. The length (several hundred microns) and diameter (1 μm or so) are not inconsistent with effective spread by passive means alone. If both "dendritic" and "axonal" arborizations have local input-output relations, however, as discussed above, then the arborizations will function independently of each other to a certain extent. It is possible, therefore, that the primate horizontal cell provides separate branching systems whose functions are not necessarily sequential, as is implied by "dendritic" and "axonal" labels, construed in their classical sense.

In lower vertebrates, the input resistance (R_N) of horizontal cells has been reported to range from 1-30 MΩ (see Witkovsky, 1971). In the fish it is clear that spread of electrical current through the horizontal cells is effective, for it has been possible, in recent experiments, to inject small amounts of current into a horizontal cell and influence thereby the activity of distant ganglion cells. This influence is mediated through electrical synapses between the horizontal cells, and synaptic pathways through bipolar cells and, possibly, amacrine cells as well. Electrotonic considerations relevant to this case have been discussed by Marmarelis and Naka (1972).

AMACRINE CELLS   Amacrine cells have a predominantly lateral orientation, resembling that of horizontal cells. They differ, however, in that they are clearly neuronal in morphology; they generate depolarizing synaptic potentials, and they appear to have limited impulse-generating properties.

The amacrine cell processes occupy both presynaptic and postsynaptic positions within the IPL [see Fig. 35(C, D)], and those processes, therefore, function as local input-output units. As in the horizontal cells, spread of synaptic potentials between branches is presumably effective and graded in intensity and in spatial extent.

The nature of the local feedback onto the bipolar cell terminal, through the reciprocal synapse (see Fig. 35), is one of the puzzling aspects of amacrine cell function. An analogy with the reciprocal synapses in the olfactory bulb has been suggested (Rall et al., 1966; Dowling and Boycott, 1966; Stell, 1972). In the olfactory bulb, the reciprocal synapses of granule cells provide for inhibitory feedback onto the mitral cells; this feedback is triggered by the initial impulse in the mitral cell, and the feedback occurs in the aftermath of the impulse. In the retina, the input from the BP terminal is a slow potential; since the amacrine cell responses tend to be transient, it has been suggested (Werblin and Dowling, 1969) that the reciprocal synapse could mediate inhibitory feedback onto the bipolar cell terminal to shut off the bipolar input after the initial excitation.

Since the amacrine cell is placed at the interface between hyperpolarizing and depolarizing elements in the retina, it raises the possibility that the amacrine cell could, through postsynaptic depolarization at all its terminals, oppose the excitatory input from the bipolar cells, on the one hand, while exciting amacrine and ganglion cells, on the other. The interface thus extends the range of possible mechanisms for amacrine cell modulation of synaptic transfer at this level. Even given these differential effects, it seems necessary to postulate different actions at some amacrine cell synapses to account for the IPSP's in ganglion cells (Fig. 37).

The amacrine cell is also faced with the problem of providing for transmission to distant parts of its branching tree. The lateral field may extend up to 1 mm in diameter in some retinas. Few estimates of electrical parameters of the amacrine cells have been made, and it is clearly premature to suggest any kind of electrotonic model (see Nelson, 1973). It would appear that attenuation might be severe through the entire extent of an amacrine cell, but the amacrine cell, nonetheless, may be assumed to provide for such spread, since the experimental evidence implicates it in the lateral processing that underlines the extensive center-surround fields of ganglion cells, as well as other properties in more complex retinas (see Dowling, 1970). It may well be that a limited amount of impulse activity assists in promoting lateral spread. This might be analogous to the role of dendritic spikes in cerebellar Purkinje cells (Chapter 8) and hippocampal pyramidal neurons (Chapter 11)

GANGLION CELLS The ganglion cell dendrites are, as far as is known, exclusively postsynaptic in position. This means that, like motoneuron dendrites, their local role is concerned with integration of synaptic inputs preparatory to transmission to the cell body and axon hillock. It is assumed that this integration takes place by passive spread; no evidence of active properties of the dendrites has been reported. The dendritic trees of the smallest ganglion cells, in the foveal region, are quite limited in extent, so it may be presumed that spread of synaptic potentials within them is very effective. This, of course, is consistent with their role of providing for the highest acuity in the visual periphery.

In complex retinas, the dendritic trees of ganglion cells have complicated patterns, as indicated in Fig. 34. The possibility that different functional classes may be correlated with different patterns of dendritic arborization was investigated by Maturana et al. (1960) in their classical study of visual processing in the frog. They obtained evidence that the number of distinct levels of stratification of dendritic branches did, in fact, provide a basis for explaining the functional classes of ganglion cell responses. The idea seems reasonable. In a recent study (West and Dowling, 1972), however, electron microscopy of Golgi impregnated ganglion cells indicated that cells with the same branching pattern may differ widely in their synaptic connections with amacrine and bipolar cells. This implies that ganglion cells of the same anatomical type have different functional properties.

These results should not be taken to mean that dendritic geometry is not important in assessing structure-function relations within the brain; it means, simply, that these relations must be assessed with caution. Particularly to be resisted is the notion that a specific function is correlated with a specific structure. To illustrate this point, consider (as is obvious) that the amacrine and ganglion cells have become differentiated in complex retinas to subserve complex functions. Does this imply a unique relation between the geometrical patterns of these retinal cells and those specific functions? It is easy to show that this cannot be the case, for, in animals (e.g., primates) with simple retinas, abstraction of visual stimuli is postponed to the cerebral cortex, an expression of the encephalization of nervous control. This means that, within the cortex, neurons that are radically different from the retinal neurons in size, shape, and branching pattern nonetheless perform steps of sensory processing similar to those performed by the retinal neurons. We will review that evidence in discussing the neocortex (Chapter 12). It must be concluded, therefore, that a flexible relation exists between structure and function in the neurons of the brain and that *similar functions may be subserved by different structures.* One may also conclude that the obverse is true: *different functions may be subserved by similar structures;* the extraordinary range of properties of neuronal dendrites may be taken as evidence of this. Recognition of these facts about structure-function relations is necessary in formulating the principles underlying the synaptic organization of different regions of the brain.

8

CEREBELLUM

The cerebellum, an outgrowth from the brain stem, is an important part of the brain that is present throughout the vertebrates. Its relative size and position can be seen in the diagram of Fig. 2. It is a largely central region, the first that we take up in our study. From its strategic position, it makes connections with many other brain regions, motor, sensory, and otherwise, and because of this it is not nearly as easy to specify its specific functions as it is for the olfactory bulb and retina, with their single afferent inputs, or the spinal ventral horn, with its single motor output.

Studies of animals with cerebellar ablations, and of humans with traumatic injuries and disease, indicate that the cerebellum is involved in at least three broad functions: (1) maintenance of posture and balance, (2) maintenance of muscle tone, and (3) coordination of voluntary movements. Although these functions express themselves as motor acts, they nonetheless also depend heavily on specific sensory inputs. Posture, for example, requires information from spindle receptors about the state of muscle contraction. Balance requires information from the vestibular canals. Coordinated movements require many additional inputs: somatosensory, visual, etc. The cerebellum must therefore be looked upon as an organ for sensori-motor coordination, not merely as a motor appendage.

Like the olfactory bulb and retina, the cerebellum is a laminated region in which the basic internal structure is similar throughout. The structure does vary considerably in vertebrate phylogeny, as has been summarized by Llinás and Hillman (1969). There are also various anatomical subdivisions of the cerebellum, which are covered in neuroanatomy textbooks. Our concern will be restricted to the lateral hemisphere in the mammal, in order to illustrate general principles of cerebellar organization.

NEURONAL ELEMENTS

The cerebellum is a highly convoluted structure, its layers being thrown into numerous folds. It resembles in this respect many nuclear regions in the brain stem, as well as the cerebral cortex of higher mammals. Convolutions in the cerebellum, as elsewhere, increase the amount and extent of the region, presumably in response to increasing demands for the functions performed therein.

Within the convolutions a rigid geometry dominates the organization to an extent not found in other central brain regions. This is an outstanding characteristic of the cerebellum and must be incorporated, along with the convolutions, in the description of the neuronal elements. In order to do this a three-dimensional diagram is commonly constructed, as in Fig. 39. This diagram preserves the same scale of magnification used in the other diagrams of this book. In most of the cerebellum the convolutions are oriented transversely to the longitudinal axis of the animal.

We follow in the main the classical account of Cajal (1911), the account of Eccles, Ito, and Szentagothai (1967), and the recent work of Palay and Chan-Palay (1973) and of Szentágothai and co-workers (Palkovitz, Magyar, and Szentágothai, 1971).

INPUTS   The cerebellum receives its input from fibers arriving through the depths of the cortex. It resembles in this respect the neocortex and differs from other "cortical" regions—olfactory bulb, olfactory cortex, and hippocampus, as well as retina—in which the afferent component of the input arrives at the outer surface.

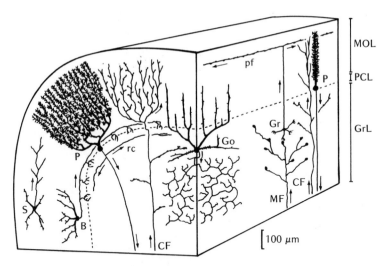

FIG. 39. Neuronal elements of the mammalian cerebellum.

Inputs: mossy fibers (MF) and climbing fibers (CF).

Principal neuron: Purkinje cell (P), with recurrent collateral (rc).

Intrinsic neurons: granule cell (Gr); stellate cell (S); basket cell (B); Golgi cell (Go).

Histological layers are shown at the right: molecular layer (MOL), Purkinje cell body layer (PCL), granule layer (GrL).

It is usual to divide the input into two types. One is the *climbing fiber*, so called by the manner in which it ascends through the cortical layers. It is thinly myelinated and has a diameter of 1-3 μm. Within the deeper, granule layer (GrL, Fig. 39), it loses its myelin and arborizes into long branches, which ascend through the cortex. Some collaterals are given off within the granule layer; most terminate within the superficial, molecular layer (MOL, Fig. 39). The branching field is very extensive across a convolution (folium) but very restricted along it; in this it resembles the dendritic tree of the Purkinje cell (see below).

The other type of input fiber is the so-called *mossy fiber*. These fibers also arrive through the depths, but they extend only into the granule layer. They are somewhat larger in diameter than the climbing fibers and much more numerous. On their way, they branch repeatedly, so that one fiber may supply several folia and several areas within one folium. Within the granule layer they lose their myelin, branch several times, and terminate in the characteristic, large, mossy-like terminals that give them their name. There is no particular orientation in the field of mossy fiber branches.

The sources of the two types of input fiber have been difficult to determine. Recent evidence indicates that the climbing fibers originate mainly in the *inferior olivary complex* (see Eccles et al., 1967), a region that, in turn, receives input from all three levels of the brain: spinal cord, brain stem, and cerebral cortex. The mossy fibers, on the other hand, appear to have much more widespread origins. The vestibular nerve, which carries sensory information about balance from the *vestibular canals*, has some axons that end in mossy terminals in a part of the cerebellum known as the flocculus. An important source is the nearby *pontine nuclei* of the brain stem, which in turn receive direct connections from the cerebral cortex. Other sources are the *spinal cord* (through the spino-cerebellar tracts), the *reticular nuclei*, and the *vestibular nuclei*. Some of these inputs show a preference for terminating at different levels within the granule layer. In addition to these input sources, the locus coeruleus has been of interest recently as part of an adrenergic brain-stem system, which may send fibers to the cerebellum (Bloom, Hoffer, and Siggins, 1971; Olson and Fuxe, 1971).

It may be noted that we have not described the inputs in terms of a division between afferent and central, or central and centrifugal, as can be done in most other regions of the brain. The vestibular and spino-cerebellar inputs, arriving through mossy fibers, are analogous to the sensory afferent inputs of other regions. The other mossy fiber inputs, and the climbing fiber inputs, would by this analogy be comparable to the central inputs of other systems.

The differences between the climbing fiber and mossy fiber terminals seem remarkable, if not unique, but it should be recalled that the afferent and central inputs to the olfactory bulb are also markedly dissimilar. It is also remarkable that so many different functional "modalities" are funneled into the cerebellum through the same rigidly stereotyped structure of the mossy fiber terminal. Despite their differences, climbing fibers and mossy fibers have some synapses onto Golgi cells and granule cells that are similar in structure. Chan-Palay and Palay (1971b) have pointed out the interesting implication that the form of a terminal is determined by the postsynaptic site.

PRINCIPAL NEURON   The output from the cerebellum is carried through the axons of the *Purkinje cells*. Although it is sometimes said that every cell of the brain is unique, the Purkinje cell is "more unique", if that were possible, than most. The cell bodies are arranged in a single sheet about 400 μm below the surface, at the junction of the molecular and granule layer. As shown in Fig. 39, each Purkinje cell has a large dendritic tree that is flattened in one plane, so that it looks like a pear tree espaliered against a garden wall. In its branching field and plane of orientation, the dendritic tree is similar to the climbing fiber branches.

The Purkinje cell bodies have diameters ranging from 20 to 40 μm. Each has a stout dendritic trunk (up to 10 μm or so in diameter) that gives rise to a sequence of primary, secondary, and tertiary branches. The branching tree stretches across the molecular layer to the surface, so that in vertical and horizontal extent it is 400 μm or so, whereas across its plane of orientation it is little more than the diameter of its branches (see Fig. 39). The smaller dendritic branches are profusely invested with *spines* or thorns; these are basically similar in outward appearance to the spines of olfactory granule cells and of pyramidal cells in the hippocampus and neocortex. It has been estimated that a single Purkinje cell has upwards of 100,000 spines. This is only the first of many amazing numerical counts we will mention that have been obtained in the analysis of cerebellar elements and circuits.

The rigidity of cerebellar geometry is clearly expressed in the Purkinje cell dendritic tree, the single sheet of cell bodies, and the relatively homogeneous size and shape of the Purkinje cells. No subgroups (as seen in olfactory mitral and tufted cells or the varieties of retinal ganglion and cortical pyramidal cells) have as yet been described. Compared to other output neurons, Purkinje cells, therefore, appear as a very stereotyped population.

The Purkinje cell axons give off numerous recurrent collaterals within the folium of origin. The branches of the collaterals form a plexus at the level of the Purkinje cell bodies. As shown in Fig. 39, the branches spread predominantly across the folium, in the plane of the Purkinje cell dendritic tree. The collaterals are myelinated up to their terminals.

The axons descend through the granule layer, wherein they gain a myelin sheath and carry the output away from the cerebellum. The remarkable feature of this output is that most of it goes to three nearby nuclei that are packed into the depths of the cerebellum; they are referred to collectively as the *deep cerebellar nuclei*. In the parts of the cerebellum with which we will be concerned, all the axons terminate in these nuclei; hence, the cerebellar output is relayed through them to the rest of the brain. The Purkinje cell axons are, therefore, relatively short, no more than a few millimeters in length in small animals.

The deep nuclei act as relays from specific parts of the cerebellum to specific parts of the brain. The newest part of the cerebellum, the neocerebellum, projects through its Purkinje cells to the dentate nucleus. In higher mammals this nucleus is deeply convoluted, in correlation with the enlarged neocerebellar hemisphere. It projects mainly to the *ventral lateral nucleus* of the thalamus and also to the *red nucleus* (a motor integration center with connections to the spinal cord) and to a part of the *reticular nuclei*. Since the dentate and other deep nuclei also receive inputs from branches of the climbing and mossy fibers on their way to the cerebellum, it is obvious that we must include the deep nuclei in studying the functional organization of the cerebellum.

INTRINSIC NEURONS   There are three main types of intrinsic neuron in the cerebellum, all distinct from the Purkinje cells and from each other.

Granule cells fill the granule layer (see Fig. 39). The cell bodies are very small (6-9 μm in diameter). Each gives rise to three to five dendrites, which are less than 1 μm in diameter and no more than 30 μm in length. The cell bodies and their dendrites are thus among the smallest in the brain. Each dendrite terminates in a claw-like expansion that is the site of synaptic connections.

The granule cell gives rise to a thin axon (1 μm or less in diameter) that ascends several hundred microns through the granule layer into the molecular layer. Here it divides, in a T-shaped fashion, to give rise to two branches that run horizontally through the molecular layer. These branches have their own name of *parallel fibers*, from the fact that they are all arranged in parallel in the axis of the folium. In this orientation they run perpendicularly through the Purkinje cell dendritic trees, as is shown in Fig. 39. The parallel fibers are unmyelinated, with diameters of only 0.2 μm near the surface of the molecular layer; these therefore resemble olfactory nerve axons. Their diameter gradual increases to about 1 μm in the depths of the molecular layer. Each parallel fiber branch is 1-1.5 mm in length, i.e., several times longer than the ascending axon from which it arises.

It should be stressed that the cerebellar granule cell differs morphologically in almost every respect from the olfactory granule cell and from the granule cells of the dentate gyrus (Chapter 11) and cerebral cortex (Chapter 12), each of which also differs radically from the others. The term "granule" has no general significance as thus employed; it only describes the dot-like appearance of the small cell bodies in the primitive preparations of the early histologists.

Golgi cells are relatively large neurons, with cell bodies about the same size as Purkinje cells. The cell bodies are scattered throughout the superficial part of the granule layer. Each gives off several large and rather straight-appearing dendrites; most of them ascend and branch in the molecular layer, although a few are directed to the granule layer (see Fig. 39). The field of arborization is unoriented and very wide, perhaps three times as wide as a Purkinje cell dendritic tree (i.e., up to 1 mm across).

There are few dendritic branches, and the branches are only sparsely invested with spines. The axon is striking in appearance. It arises from the cell body or a dendritic trunk and immediately, within a few microns, begins dividing repeatedly, giving rise to a dense arborization of short branches that span the entire granule layer. It is surely one of the most fantastic arborizations of any axon in the brain. In breadth, the field of branching approximates the extent of the dendritic tree. The branches end in clusters of terminals, which resemble the claw-like terminals of the granule cell dendrites.

In the molecular layer are several types of *stellate cells*, so called because of the star shape of their dendritic trees. The most highly differentiated type is the *basket cell*. These cell bodies are situated in the deeper part of the molecular layer. Each is some 20 μm in diameter. The dendritic tree extends for 100-200 μm. The branches bear spines, and the tree is oriented across the folium, much as the Purkinje cell dendrites are. Each cell gives rise to an axon, 1-2 μm in diameter, that is oriented across the folium. During its course it gives off branches, which entwine the Purkinje cell bodies in a basket-like form, which gives these cells their name (see Fig. 39). A single basket cell may give rise to as many as ten baskets, to as many Purkinje cells, in its course over a distance of 1 mm or so. The terminals envelope the cell body and also the initial segment of the Purkinje cell axon, making contacts that are described in the next section.

Other varieties of stellate cell are found more superficially in the molecular layer. The cell bodies and dendritic trees are generally smaller than those of the basket cells. The axons tend to run in the same horizontal plane, but the branches are simple and short and oriented vertically (see Fig. 39). There is a gradation of transitional forms, from the basket cells to the simplest and smallest stellate cells near the surface of the molecular layer.

It may be noted that none of these intrinsic neurons strictly satisfies the classical definition of a short-axon cell, that is, a cell whose axon ramifies in the vicinity of the cell body. On the other hand, they do satisfy our previous definition (Chapter 6) of a cell whose axon distributes within the same histological region (i.e., the cerebellum). But within this very general classification, the cerebellar short-axon cells are markedly different in form and function from other short-axon cells and from each other. We will see that the granule cell is actually a type of relay neuron in the input pathway from mossy fibers to Purkinje cells. The Golgi cell, with its distinctive dendrites and fantastically arborized axon, is a unique type of intrinsic neuron. Only the stellate cells of the molecular layer are recognizable as short-axon cells that resemble their counterparts elsewhere in the brain.

CELL POPULATIONS The cerebellum has been a favorite subject for the geometricians and arithmeticians of the brain. If the geometry is striking, the arithmetic is no less than astounding. Rolled out into a flat sheet, the cortex of a cerebellar hemisphere of man would cover a surface approximately 2×100 cm (see Braitenberg and Atwood, 1958). This is some two-hundred times the area of the retina, and almost one-third the area of a hemisphere of the entire cerebral cortex. Within the cortex of a cerebellar hemisphere are approximately seven-million output neurons (the Purkinje cells); this is seven times the estimate of retinal ganglion cells in man, and more than one-hundred times the estimate of olfactory mitral cells. From these numbers alone one would anticipate that the cerebellum must provide a dominant input to the brain. Recalling our discussion of neuronal populations in Chapter 7, would one say that we are "exceptionally cerebellar animals"?

The populations of input fibers are no less remarkable. There is a roughly one-to-one ratio of climbing fibers to Purkinje cells. Because of this, there is little or no convergence or divergence in this input pathway. The mossy fibers, on the other hand, are far more numerous. No figures are available, but they must outnumber the climbing fibers by several orders of magnitude, which puts their numbers in the billion range. Compared with the figures mentioned in Chapter 7, this is far more input than any sensory system provides. It thus appears that the two types of cerebellar inputs stand at extremes, one at the limiting minimum of convergence and divergence, the other at the maximum.

It is in the population of granule cells that the arithmeticians come into their own. Consider the estimate of 2.4 million cells/mm³ in man (Fox and Barnard, 1957); this works out to be approximately 20 billion granule cells in one cerebellar hemisphere, or 40 billion in both hemispheres. It is commonly stated that the human brain contains 10 billion nerve cells, as evidence of its fantastic complexity; clearly those statements do not take into account the granule cells of the cerebellum!

The other two types of intrinsic neuron are much less numerous; in fact, they tend to be scarce or absent in lower vertebrate species. In mammals, the ratio of Golgi to Purkinje cells is only about 1:10, whereas the numbers of stellate and basket cells are probably of the same order, or somewhat greater, as the Purkinje cells. If we consider the granule cell to be a relay neuron, then the populations of intrinsic neurons, i.e., Golgi, stellate, and basket cells, can be seen to be quite modest in number. The dendritic trees and extensive axonal arborizations of these cells mean that their convergence and divergence ratios vis-à-vis the Purkinje cells are much greater than the numbers of cell bodies would indicate, however. For a detailed study of neuronal populations, and convergence and divergence ratios, in the cat, the reader is referred to a recent study of Palkovits et al. (1971) (summarized in Eccles, 1973).

This brief account will serve to indicate that the cerebellum is characterized by extraordinarily large populations of some of its neuronal elements, which are, in addition, packed in together at a very high density. The significance of these population figures depends on an accurate knowledge of the internal circuits of the cerebellum. Further quantitative data will, therefore, be mentioned in conjunction with the basic circuit diagram.

SYNAPTIC CONNECTIONS

In the cerebellum, as in the olfactory bulb and retina, the distinct cell types and lamination greatly expedite the identification of

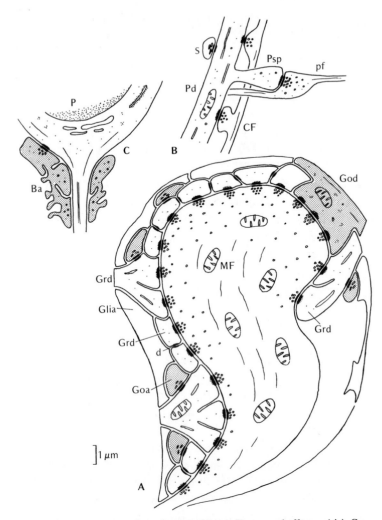

FIG. 40. Synaptic connections in the mammalian cerebellum. (A) Synaptic glomerulus in the granule layer, showing axodendritic connections between a mossy terminal (MF), granule dendrites (Grd), and Golgi dendrites (God); also from Golgi axons (Goa) onto granule dendrites. Note the desmosomes between the granule dendritic terminals; note also the surrounding glia. (B) Molecular layer, showing axodendritic connections from parallel fibers (pf), climbing fibers (CF), and stellate cells (S) onto a Purkinje cell dendrite (Pd) and dendritic spines (Psp). (C) Specialized terminal of a basket cell axon (Ba) onto the cell body, axon hillock, and initial axonal segment of a Purkinje cell (P). (After Eccles et al., 1967.)

synaptic connections. Indeed, as in the olfactory bulb and the retina, the main types of connection were inferred from Golgi stained preparations by the classical histologists of the late nineteenth century (see Cajal, 1911). Studies with the electron microscope have confirmed most of these inferences and revealed important details of synaptic structure (see Eccles et al., 1967; Llinás and Hillman, 1969; Mugnaini, 1970; Rakić and Sidman, 1973; Palay and Chan-Palay, 1973).

We begin with the afferent input, and consider the connections made by the mossy fibers. As shown in Fig. 40, the mossy terminals, as viewed under the electron microscope, appear as enlarged swellings, termed *rosettes*. The rosette is one of the largest terminal structures in the brain; the medium-sized example shown in Fig. 40 is some 10 μm in width and 20 μm in length. This terminal structure is, therefore, larger than the cell bodies of most small neurons in the brain.

Tightly grouped around a rosette are numerous terminals, the whole cluster being surrounded by a single layer of glial membranes. The glial membranes thus demarcate a specific group of terminals, which is termed a *glomerulus*. Within a glomerulus, the mossy rosette makes synaptic connections onto dendritic terminals of granule cells, as shown in Fig. 40. These are type I chemical synapses. The rosette also has synapses of a distinctive type onto Golgi cells; because the surface of the Golgi cell is wrinkled like a Spanish chestnut, these synapses have been termed *en marron* (Chan-Palay and Palay, 1971b). A third type of connection within the glomerulus is from axon terminals of Golgi cells onto granule cell dendrites; these are more characteristic of type II chemical synapses. In many cases, a granule cell dendritic terminal receives a connection from both a mossy rosette and a Golgi axon. The granule dendritic terminals are interconnected by desmosome-like membrane specializations. It has been estimated that a single glomerulus contains 100 to 300 dendritic terminals from some 20 granule cells.

The term *glomerulus* has already been used to describe the large areas of neuropil in the olfactory bulb, and we will encounter it again in our description of the organization of the thalamus (Chapter 9) and the cerebral cortex (Chapter 12). It is, therefore, one of those terms likely to give rise to confusion, and it is appropriate to consider it more carefully. What we are basically dealing with in these cases is a *synaptic complex*, which may be defined (see Shepherd, 1972a), in the most general sense, as *a set of specific synaptic connections between axonal and dendritic terminals, which terminals may themselves be interconnected.* This generalizes previous definitions (see Szentagothai, 1970; Pinching and Powell, 1971) to cases of dendrodendritic as well as axodendritic connections within a synaptic complex. *Glomerulus* may then be defined, again in the most general sense, as *a synaptic complex enclosed in glial membranes or otherwise set apart.* The olfactory glomerulus, involving many axonal and dendritic terminals, might appropriately be termed a *macroglomerulus*; this term might also apply to the barrels in the neocortex (see Chapter 12). The cerebellar and thalamic type, on the other hand, involving only a small number of terminals, could be termed a *microglomerulus*. The synaptic complex formed in relation to single olfactory axon terminals, as described in Chapter 6, is analogous to the synaptic complex of the cerebellar and thalamic microglomeruli, although it lacks the well-defined glial enclosures of the latter. The synaptic patterns within the cerebellar glomerulus will be compared with these other cases when we describe the basic circuit, below; a detailed comparison with the thalamic glomeruli is provided in Chapter 9.

Within the molecular layer of the cerebellum are three major types of connection onto the Purkinje cell dendrites. First, the parallel fibers, arising from the granule cell axons, have synapses onto the spines of the Purkinje cell dendrites, as is shown in Fig. 40. These connections are type I chemical synapses. The presynaptic process is simply an enlargement of the parallel fiber as it passes by the spine; this is an *en passage* or a *crossing-over* synapse. This is the only type of synapse made by the parallel fibers; such synapses are also made onto the dendritic spines of the basket, stellate, and Golgi cells in the molecular layer.

The second major type of connection onto the Purkinje cell dendrites is made by the climbing fibers. This type also may be regarded as a variety of *en passage* synapse, but at right angles to the parallel fibers, as the climbing fiber ascends along the Purkinje cell dendrites. These are also type I chemical synapses and are exclusively to the trunks rather than the branchlets of Purkinje cell dendrites; thus, these two types of input are highly specific for different sites on the Purkinje cell dendritic tree.

A third type of connection is made by the axon terminals of stellate cells onto the surfaces of the Purkinje cell dendrites. These are type II chemical synapses (see Fig. 40).

A distinctive type of synaptic connection is found near the Purkinje cell axon hillock. As we know, the basket cell axon entwines the Purkinje cell body. Electron microscopy shows that the terminals are of unusual design, as illustrated in Fig. 40 (C) (see Eccles et al., 1967). On the cell body near the axon hillock, the basket cell terminals establish contacts which generally fall into the category of type II chemical synapses. In addition, further terminals which surround the axon hillock and initial segment are given off. These contain synaptic vesicles, but there is little specialization of the apposed membranes. The terminals have numerous villi-like protuberances, which interdigitate with glial cells (the Bergman glia). Thus is formed a cap around the axon hillock region of the Purkinje cell, reminiscent of the cap around the corresponding region of the giant motor cells (Mauthner cells) of the fish spinal cord.

Two characteristics of the synaptic connections in the cerebellum are especially significant in comparison with other brain regions. One is that there are synapses from recurrent collaterals of the Purkinje cell axons onto the cell bodies and dendritic trunks of neighboring Purkinje cells. The presynaptic terminals contain flat and round synaptic vesicles, but the synapses have been reported not to conform to either of Gray's types (Chan-Palay, 1971). Such recurrent connections of an output neuron onto itself, as well as onto intrinsic neurons, have also been reported in certain parts of the cerebral cortex (see Chapters 10 through 12). We will discuss further the general significance of this type of connection in those chapters. In other regions of the brain, the recurrent collaterals of the output neuron synapse only on the intrinsic neurons; examples are the olfactory bulb, the thalamus, and, probably, the ventral horn.

The other characteristic is that all the synapses thus far described are of the axosomatic or axodendritic type. No dendro-dendritic synapses of the chemical type have been described, although there are densities of the opposed membranes of granule cell dendrites within the glomerulus (see Fig. 40). Nor have axoaxonic connections been observed, with the exception of the special type formed by the basket cell terminals onto the initial segment of the Purkinje cell. This limited stock of synaptic types is in contrast to the greater variety of connections in many other parts of the brain.

BASIC CIRCUIT

The synaptic organization of the cerebellum is summarized in the basic circuit diagram of Fig. 41. The first diagram (A) preserves the topographical orientation of the cerebellum and includes as much of the internal circuits as is feasible. Diagrams (B) and (C) are oversimplified rearrangements of the neuronal elements in order to emphasize aspects that are of interest compared to the organization of other regions in the brain.

In general, it may be said that the main plan of the cerebellum is laid down by the two input pathways and their relations to the output neuron. This plan is emphasized in the simplified diagram of Fig. 41(B, C), in which the inputs are shown arriving from above, in order to facilitate comparison with the input-output connections in other basic circuit diagrams. If we take first the climbing fibers [cf. Fig. 41(A)], they make synapses directly onto the Purkinje cells; as we shall see, this is an excitatory pathway. This is therefore analogous to the monosynaptic input to motoneurons, and the olfactory input to mitral cells, in having direct access to the output neuron. The climbing fibers also make

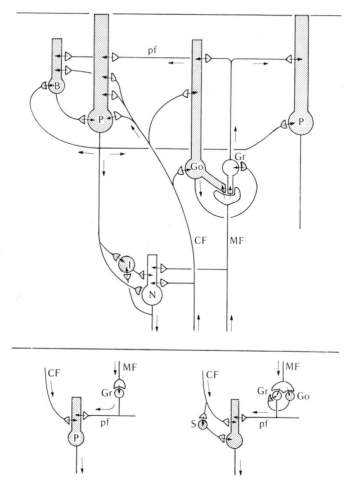

FIG. 41. Basic circuit diagram for the mammalian cerebellum. Abbreviations as in Fig. 40. Note deep cerebellar nuclear cells, including the principal nuclear cell (N) and intrinsic neuron (I). *Below*, simplified diagrams to emphasize the two main input pathways in the cerebellum, for comparison with other brain regions. *Right*, these pathways shown with their associated intrinsic neurons.

connections onto the stellate (S) and basket cells [B, Fig. 41 (A, C)], the intrinsic neurons of the molecular layer that control the Purkinje cell output. We thus recognize the basic elements of a synaptic triad in this pattern of connections. The intrinsic neurons are differentiated into two distinct subtypes; one of these types—the basket cell—receives axon collaterals from the output neuron it controls; this recurrent pathway is similar to pathways in the ventral horn and olfactory bulb.

The mossy fiber input to the Purkinje cells is an indirect one, being relayed through the granule cells (Gr); this is analogous to the disynaptic input to motoneurons in the ventral horn. The granule cell functions, therefore, as a relay neuron in the vertical path onto the Purkinje cells. As will be shown later, the mossy fiber is excitatory to the granule cell, and the granule cell (through its parallel fiber) is excitatory to the Purkinje cells. Thus, the granule cell is not interjected into the vertical pathway to convert an excitatory to an inhibitory input, as in the case of many interneurons in the spinal cord, for example. In this sense, the granule cell is a true relay neuron (cf. Cajal, 1911; Llinás and Hillman, 1969) rather than a type of interneuron or short-axon cell for local processing. It may be characterized as a local, in contrast to a long-distance, output neuron.

This view is supported by the fact that in certain animals (e.g., electric fish) the granule cells do not lie in a layer within the cerebellar cortex but instead are clustered as a nucleus of cells at the base of a folium (ridge) (Nieuwenhuys and Nicholson, 1969). This arrangement suggests an essentially nuclear function for the population of granule cells and further shows that an ostensibly cortical structure, such as the cerebellum, may contain neurons that have a nuclear type of organization. The implication here is that the molecular layer is the true cortical part of the cerebellum and that the functional demands on the parallel fiber relay in higher vertebrates require the granule cells to be as close as possible to the molecular layer. The structure of the molecular layer is nonrepeating in relation to the output neuron (Purkinje cell) that spreads across it, in line with our previous discussion of one of the principles of cortical organization.

Seen in this light, the granule cells form what may be termed a *staging* or *pre-processing station* for input to the cerebellar cortex. This position can be appreciated more fully in the simplified diagrams of Fig. 41(B, C), in which the mossy-granule cell relay has been reoriented to feed into the cerebellum as a separate entity from above. In this position, it can be compared with the interneuronal relays in the spinal cord (cf. Fig. 19, Chapter 5) and with the dentate relay to the hippocampus (Fig. 58, Chapter 11); it can also be compared with the thalamic relay to the neocortex, to be discussed in Chapter 9.

Let us now consider more closely the synaptic arrangements in the mossy fiber input pathway. The excitatory connections, as already described, are found in all vertebrates. The additional connections made by intrinsic neurons provide the basis for the more complex cerebellums of higher vertebrates.

In the granule layer, the mossy fiber rosette makes synapses onto both the granule cells and the Golgi cells (GO) as shown in Figs. 40 and 41. We recognize here the basic elements of a synaptic triad; the input fiber (rosette), the output neuron (granule cell), and the intrinsic neuron (Golgi cell). The basic pattern is one of simultaneous input to the dendrites of both output and intrinsic neurons, but with no connections between the dendrites of those neurons. The cerebellar glomerulus is thus a synaptic complex which provides for a very simple stereotyped pattern of connections within the synaptic triad. In Chapter 9, we will compare this pattern of organization with the synaptic triads in the olfactory bulb, retina, and thalamus.

As can be seen in Fig. 41(A, C), Golgi cells function as intrinsic neurons in controlling input-output relations between the mossy terminals and the granule layer branches; thus, there are both local circuits (within the granule layer) and longer circuits (through the molecular layer) for feedforward and feedback inhibition. Through these circuits, the Golgi cell regulates the granule cell relay to the Purkinje cells. With regard to the Purkinje cells, then, the Golgi cells are at the level of input processing.

In the molecular layer, the parallel fibers of the granule cells make connections onto the dendrites of Purkinje cells and also onto the dendrites of basket and stellate cells. The basic pattern of this synaptic triad is a rigid, near-simultaneous sequencing of input to both output and intrinsic neurons, with no dendrodendritic interactions between the latter. It is especially significant that the output and intrinsic elements in this synaptic triad are the same as those for the climbing fiber input, although the pattern of input connections to them are distinctly different.

An understanding of the functional significance of the pathways within the cerebellum depends on the quantitative data. If we put together the estimates that a single mossy fiber has 40 rosettes, that each rosette connects to the dendritic terminals of 20 granule cells, and that a single granule cell (through its two parallel fiber branches) connects to 100-300 Purkinje cells, we obtain a combined divergence ratio of roughly 1:100,000-300,000 in going from one mossy input fiber to the Purkinje cell. The Purkinje cells, for their part, have upwards of 100,000 dendritic spines, each spine with its parallel fiber synapse onto it, and this is, therefore, roughly the number of parallel fibers, and, hence, individual granule cells, that converge onto it.

These factors for convergence and divergence between input and output are greater than those known for any other region of the brain. It is as if the extreme limits of these factors had been explored, and the mossy fiber-granule cell-parallel fiber-Purkinje cell system devised as an answer. The implications of this system are that the input arriving over an individual mossy fiber is disseminated to a population of granule cells, that the activity relayed by this granule cell population passes in a "beam" of parallel fibers to the Purkinje cells, and that this beam (spreading in opposite directions from the origin of the two parallel fiber branches) activates the Purkinje cells in a rigid temporal sequence. The input from one mossy fiber must be tiny relative to the total convergence onto a given Purkinje cell. An important consequence of this last point is that summation of granule cell inputs must be a requirement for eliciting responses over this route.

If the Purkinje cells are thus buffered, as it were, from the mossy fiber input, they are, by contrast, exposed as directly as possible to the climbing fiber input. As already noted, the convergence and divergence factors for this input are close to one. In addition, a single climbing fiber has perhaps several hundred synapses onto the single Purkinje cell over which it climbs. It is as if, in the face of the massive mossy fiber system, the climbing fibers had been devised to provide a means by which a single afferent fiber could have the most secure and effective means of eliciting a response in a Purkinje cell. We will see that this inference from anatomical considerations is supported by the extraordinary potency of this synaptic connection.

It should again be emphasized that the connections summarized in Fig. 41 occur within a rigidly geometrical framework. The horizontal and vertical aspects of organization that we have pointed out in the basic circuits for the olfactory bulb and retina are carried to a high degree of perfection in the lattice-like arrangement of the cerebellum. From this organization emerges the interesting fact that, when we trace activity through any of the pathways within the cerebellum, there is the clear implication that the temporal sequences are closely determined by the spatial geometry. Let us quote from Braitenburg and Atwood (1958), two of the pioneers of quantitative cerebellar studies:

> The morphological evidence is strongly suggestive that one of the main properties of the cerebellar cortex might be the transformation of spatial into temporal patterns and vice versa. . . . Activity arising at one moment in a vertical cross-section of the molecular layer will reach different Purkinje-cell trees after different time intervals, depending on their distance. Conversely, the arrival of such a front of activity at any one Purkinje cell implies events in different loci at different times in the past, depending on the distance. Basically, therefore, equations relating certain input patterns to certain output patterns would be expected to contain time and distance interchangeably.

This should not be taken to imply that space and time are not also interrelated in other neuronal systems; it implies only that they are interrelated to a particularly rigid and stereotyped degree in the populations of neurons within the cerebellum. We will discuss similar examples of this type of organization in the olfactory cortex (Chapter 10) and hippocampus (Chapter 11).

DEEP CEREBELLAR NUCLEI It has already been mentioned that the Purkinje cell output from the cerebellum is directed to the nearby deep cerebellar nuclei. This is indicated in Fig. 41(A). One of the important findings of recent years has been the discovery that this input to the deep nuclei is inhibitory; this will be discussed further in the next section. Another important finding has been that both the mossy and climbing fibers give off collaterals to the deep nuclei. The deep nuclei contain populations of output and intrinsic neurons that form synaptic triads with the inputs from the cerebellum and from the other parts of the brain (Angaut and Sotelo, 1973; Chan-Palay, 1973); most of these details have had to be omitted from Fig. 41(A) for the sake of simplicity.

SYNAPTIC ACTIONS

Physiological analysis of synaptic actions in the cerebellum has been carried out using several methods: single volleys set up by an electrical shock (Granit and Phillips, 1956; Eccles et al., 1967); natural stimulation of sensory afferent pathways (Murphy, Mac-Kay, and Johnson, 1973; Eccles, 1973); and single neuron activity monitored during movements performed by awake animals (Thach, 1970). We will describe the results obtained by the use of single volleys and indicate briefly their relevance to the natural control of movement.

Let us first consider the synaptic actions of a *parallel fiber volley*, set up by an electrical shock delivered to the surface of the cerebellum. At this surface location, the electrode stimulates a narrow "beam" of parallel fibers that passes through the folium. The response of a Purkinje cell, recorded intracellularly, is shown in Fig. 42 (P′, column MF-PF). The response consists of a brief EPSP, lasting some 5-10 msec, ascribed to the beam of active

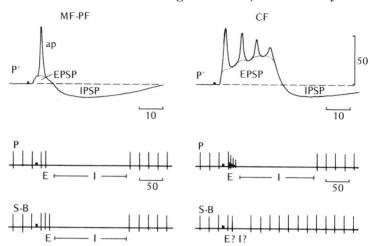

FIG. 42. Main types of synaptic actions in the mammalian cerebellum. *Left*, MF-PF, synaptic actions elicited by a volley in the mossy fiber-parallel fiber pathway. P′, intracellular recording of a simple response in a Purkinje cell. P, extracellular recording, slower time base. Excitatory and inhibitory periods indicated by E and I, respectively. S-B, extracellular recording from stellate or basket cell.

Right, CF, synaptic actions elicited by a volley in the climbing fiber pathway. P′ and P, a complex response in a Purkinje cell. S-B, ? response in stellate or basket cell. (After Granit and Phillips, 1956; Eccles et al., 1967; Thach, 1967.)

parallel fibers making excitatory synapses onto the dendritic spines of the Purkinje cell. This brief EPSP, in turn, generates a single impulse, or sometimes two or three. The response to input from the parallel fibers has been termed a simple spike (Thach, 1967).

The EPSP evoked by the parallel fibers is usually followed by a long-lasting hyperpolarization, as shown in Fig. 42. This has been described as a "prolonged, graded and chloride-sensitive hyperpolarization—the necessary and sufficient criterion for synaptic inhibition" (Llinás and Nicholson, 1971). In these respects, it resembles the IPSP's recorded from other output neurons in the brain. The entire depolarizing-hyperpolarizing sequence is, in fact, a characteristic response of principal neurons in different brain regions to a single input volley, as is pointed out in other chapters.

In other regions of the brain, IPSP's are due to feedforward and feedback connections of the intrinsic neurons (see Fig. 24). In the cerebellum it is believed, largely on inference from anatomy, that the Purkinje cell IPSP is mainly caused by a feedforward pathway through the stellate and basket cells. These cells are, indeed, activated by the parallel fiber volley; they respond with a discharge of spikes as is shown in Fig. 42 (SA, left column). The prolonged spike discharge to a single volley implies a prolonged excitatory action on these cells, suggesting that special factors (slow buildup, sequestration of transmitter, temporal dispersion) control the time course of transmitter action at these synapses from the parallel fibers (Eccles et al., 1967).

With regard to feedback connections, we have seen examples of this in the Renshaw circuit of the spinal cord (Chapter 5) and the dendrodendritic recurrent pathway in the olfactory bulb (Chapter 6). The situation is different in the cerebellum, by virtue of the fact that the synaptic actions of the Purkinje cells are inhibitory. This has been shown for the output of the Purkinje cells to the deep cerebellar nuclei (see below), and, largely by inference from Dale's Law, it has been assumed that the Purkinje cell axon collaterals are also inhibitory to the stellate and basket cells within the cerebellar cortex. Thus, rather than exciting the intrinsic neurons through recurrent collaterals, the Purkinje cells are presumed to inhibit them; since the intrinsic neurons are inhibitory to the Purkinje cells, the action of the recurrent pathway is to disinhibit the Purkinje cells, rather than to inhibit them. An inhibitory feedback pathway also exists through the recurrent collaterals back onto the Purkinje cells themselves, so it is obvious that the control of Purkinje cell excitability is the outcome of a very complex process of inhibitory interactions.

The synaptic actions elicited by a *mossy fiber volley* are similar in basic outline to those described above. The actions are neither as strong nor as limited in space, as would be expected from their dispersion by the relay through the large population of granule cells. The physiological evidence is consistent with the scheme in which the mossy rosette is excitatory to the granule and Golgi cells, and the latter are inhibitory to the granule cells. The Golgi cells thus function as inhibitory interneurons at the granule level, providing feedforward and feedback pathways as described for the basic circuit in the previous section. From the widely ramifying axonal plexi of the Golgi cells, it would appear that each Golgi cell creates a compartment of inhibited granule cells; these compartments apparently do not overlap (see Eccles et al., 1967). It is possible that the compartments are analogous to the glomeruli of the olfactory bulb (Chapter 6) and the barrels in the neocortex (Chapter 12), in the sense of constituting subdivisions of neuropil with a structural and functional unity. The term macroglomeruli has been discussed above as applicable to such compartments.

Let us turn now to the action of a *climbing fiber volley* set up by an electrical shock to the inferior olivary nucleus. The Purkinje cell response consists of an intense and prolonged depolarization, which generates a large initial spike followed by several small ones (see Fig. 42, P', column CF). This was originally termed an *inactivation response* by its discoverers (Granit and Phillips, 1956); it has also been termed a complex spike (Thach, 1967), in contrast to the simple spike response to mossy fiber input. This response is similar to the inactivation bursts of hippocampal pyramidal cells (see Chapter 11).

It is important, in this regard, to recognize that there is no one mechanism for the generation of these and other high-frequency impulse discharges in central neurons. Analysis must be carried out on each type of neuron in which it is encountered. The mechanism of these discharges is one of the most interesting challenges to the physiologist. There is evidence that cyclic AMP is an intermediary in the chemical transmission of some climbing fiber responses (Bloom, Hoffer, and Siggins, 1971).

The mechanism for the complex burst in the Purkinje cell was first explained in the following way (see Eccles et al., 1967). The volley of impulses not only propagates through the climbing fibers but also invades their axon collaterals within the inferior olive. Through these collaterals, a sequence of activation of inferior olive cells takes place, which provides a rapid series of impulses in the climbing fiber and, consequently, a rapid sequence of EPSP's in the Purkinje cells. More recently, intracellular recordings from the inferior olive cells have shown that these cells respond with a prolonged depolarization and burst of impulses, rather similar to the complex spike of the Purkinje cell. In these studies, there is evidence that the prolonged depolarization occurs in the dendrites of these cells and that it is this depolarization that underlies the brief burst of impulses in the climbing fibers and, hence, in the Purkinje cells (Crill, 1970). Recently, it has been shown that gap junctions exist between inferior olive cells (Llinás, Baker, and Sotelo, 1973), so that electrical interactions may also play some role in the discharges of these cells. There is also evidence that the prolonged depolarization of the Purkinje cell may be intrinsic, in part, to the membrane of the Purkinje cell itself (Fujita, 1969).

Whatever the outcome of these analyses, two interesting points emerge. One is that the climbing fiber synapse is extremely "potent", a physiological property that correlates with the dense innervation of the Purkinje cell by the climbing fiber. The other is that the inferior olivary cell is, to some extent, matched in its physiological properties to the Purkinje cell.

Like the simple spike response, the complex spike is also followed by a prolonged inhibition, which takes the form of a long-lasting hyperpolarizing IPSP, as shown in the intracellular recording of Fig. 42 (P′, column CF). In P, below, is an extracellular recording, on a slow time base, that shows the excitatory-inhibitory sequence as it is interposed in the resting background discharge.

The IPSP following the complex spike has been attributed to the action of local circuits. In addition to its connections to the Purkinje cells, the climbing fiber also activates stellate and basket cells, which provide feedforward inhibition to the Purkinje cells. The climbing fibers also activate to some extent the Golgi cells, which may inhibit, in turn, the granule cell excitatory input. Some of the suppression following the complex spike may be an expression of the intrinsic properties of the Purkinje cell membrane. After their initial excitation, the stellate and basket cells are themselves inhibited, presumably through Purkinje cell axon collaterals (see Fig. 41). These collaterals probably provide for similar inhibitory effects on the granule and Golgi cells (see Fig. 41).

If we take an overview of these synaptic actions, an especially striking aspect is that, whereas the actions of the input fibers (including the granule cell-parallel fibers) are all excitatory, the actions of the cells within the cerebellum are all inhibitory. This has the important consequence, as pointed out by Eccles et al. (1967), that sustained activity in response to an input is not possible within the cerebellar circuits. There are no pathways that provide for re-excitatory or reverberating activity, as have been described for the spinal cord or the cerebral cortex. Some complex sequences involving disinhibition (e.g., through Purkinje cell axon collaterals onto the inhibitory neurons) can take place, but these are apparently not sustained for long.

This aspect takes on further significance with respect to an outstanding characteristic of the cerebellar neurons, that they all have high rates of resting impulse discharge. In awake monkeys, for example, it has been found that Purkinje cells in the anterior lobe discharge at rates of 3 to 125 impulses/sec, with a mean discharge rate of 70 impulses/sec (Thach, 1972). This is in sharp contrast to the low rates, in the range of a few impulses per second, that are characteristic of output neurons in many other regions of the brain; motoneurons, as already noted, are, in fact, often silent in the absence of external input. The activity of Purkinje cells is closely correlated with the mossy fibers, which have similar high rates of resting discharge (see Jansen, Rudjord, and Walløe, 1970; Eccles, 1973). The high excitability of Purkinje cells thus derives from the synaptic drive to them, rather than from the "size principle" deduced from motoneurons (Chapter 5). Similar considerations apply to the activity of intrinsic cerebellar neurons.

It is important to realize that the synaptic actions of cerebellar inputs take place against this high-frequency background. As a consequence, both the excitatory and inhibitory aspects of the responses are significant. For example, a slight increase in resting discharge of a Purkinje cell engendered by a mossy fiber input might be difficult to detect, but an inhibitory pause following that input might be quite obvious. The extreme, high-frequency burst elicited by the climbing fiber, followed by suppression, reflects the specialization of this input for eliciting a clearly detectable Purkinje cell response no matter what the resting background discharge; at higher discharge levels, the inhibitory phase may well be the more detectable and significant part of the response.

The cerebellum and the retina provide an interesting comparison. Retinal processing is carried out almost exclusively within the domain of graded potentials, whereas cerebellar processing is carried out largely in the frequency domain. The fractionation of cerebellar circuits into an immense number of pathways, combined with the fractionation of temporal sequences into very brief intervals, appear to be the means by which the cerebellum, operating in the digital mode, achieves finely graded input-output relations like those of the retina, operating in the analog mode. Perhaps the fact that the cerebellum operates in the frequency domain reflects its close association with motoneuronal output, which, perforce, must be carried in a frequency code by impulses in the motoneuron axons to the muscles. Another close similarity is that both regions process information against a background of steady activity; the transient responses, therefore, are in the form of perturbations of ongoing activity. This apparently is a more precise mode of information transfer than is transmission by excitatory responses against little or no background. Since background activity may be adjusted under different behavioral conditions, information may be transmitted about steady states, and transient inputs are interpreted relative to those states.

DEEP CEREBELLAR NUCLEI It has already been noted in discussing the basic circuit of Fig. 41 that the cerebellar cortical output carried by the Purkinje cells is entirely directed to the deep nuclei. The work of Ito and his colleagues (Eccles et al., 1967) established that this input is inhibitory; the Purkinje cells have inhibitory synapses on the deep nuclear cells, which respond with IPSP's. The Purkinje cells are the only output neurons we will study that have an inhibitory output, and it means that the cerebellar cortical output has to be viewed through inverted spectacles, so to speak, in assessing its significance. By virtue of this output it is possible to regard the Purkinje cell as a species of interneuron, which has become displaced and has acquired elaborate internal circuits to control its own input-output relations.

The significance of the Purkinje cell input to the deep nuclei must also be assessed relative to the other inputs to the deep nuclei. A review of the synaptic organization of the deep nuclei is beyond our present scope, but anatomical and physiological studies are consistent with the following picture: both mossy and climbing fibers have connections onto both the principal and intrinsic neurons, and these connections are excitatory. Opposed to these are the inhibitory inputs from the Purkinje cells and from intrinsic nuclear neurons. Figure 41(A) gives an outline of these connections, from which some impression of the dynamic control of the nuclear output may be obtained.

The deep nuclear cells are matched with the Purkinje cells in having high rates of resting impulse discharge, ranging from 1-87/sec, with a mean of 37/sec in awake monkeys (Thach, 1972). Thus, as in the Purkinje cells, suppression of the resting discharge (by Purkinje cell input) is especially significant.

As more has been learned about the cerebellum and the deep nuclear cells, it has become clear that the relations between them are quite complex. One view that has emerged, consistent with the inhibiting function of the Purkinje cells as discussed above, is that of the cerebellar cortex as a massive accessory processing apparatus superimposed on the input-output relations of the nuclear cells with the rest of the brain. Thach (1972) has pointed out the possibility that "a step increase in the mossy fiber input might cause first a high frequency burst in the nuclear cell, and second a lowering of nuclear cell frequency as the inhibitory restraint built up in the slower Purkinje cell loop." By this means, a differentiated, rate-sensitive output from the nuclear cells could be achieved that would be similar to the kind of dynamic differentiation that is characteristic of rate-sensitive receptor cells (cf. Ottoson and Shepherd, 1971). Such rate-dependent transients might be important in the initiation of muscle movements or the control of rapid movements. This is clearly not the only possible function of the cerebellar control of the deep nuclei, and the elucidation of these functions is a most exciting challenge for future work (cf. Eccles, 1973).

DENDRITIC PROPERTIES

Let us confine our attention to the Purkinje cell. We may begin by noting that all the evidence to date indicates that the Purkinje cell dendrites are exclusively postsynaptic in position. This means that the properties of the dendrites will be relevant to local integration of synaptic inputs and transfer of the potentials to the axon hillock.

The assessment of these functions in relation to dendritic properties must be built on rigorous models for dendritic electrotonus; this was established as a guiding principle in Chapter 4, and the development of models for electrotonic current spread in motoneurons, olfactory mitral and granule cells, and visual receptor cells has been covered in the preceding chapters. This unfortunately exhausts the present catalog of neurons for which models are available; for the Purkinje cell, as for the other neurons of our study, we are restricted to identifying the questions of primary interest and summarizing the present consensus of interpretations; because of the lack of rigorous underpinning, these are more in the way of informed speculations.

ACTIVE PROPERTIES The Purkinje cell has been a focus of interest with respect to the question of active impulse properties of dendrites. The first studies were made in tissue cultures, in which

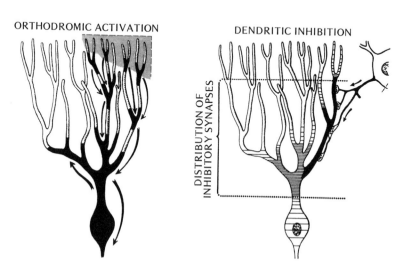

FIG. 43. Properties of Purkinje cell dendrites. (A)-(C) Microelectrode recording from within a dendritic trunk, showing responses to a parallel fiber volley of increasing strength. Note "dendritic" potentials as small spikes and bumps (dots). *Below,* schematic diagrams to illustrate *orthodromic activation* of dendritic spikes at dendritic branch points and *dendritic inhibition* by stellate cells (sc). (From Llinás and Nicholson, 1971.)

growing Purkinje cells send out long and relatively undifferentiated "dendritic" processes. With a stimulating microelectrode at one end of a process, and a recording microelectrode at the other end, evidence was obtained for a spike-like response that spread like an impulse through the dendrite (Hild and Tasaki, 1962). Such *in vitro* studies have great potential for the analysis of neuronal properties; but the fact that the neurons are embryonic, poorly differentiated, and essentially deafferented suggests the need for great caution in extrapolating the results to the adult *in vivo* situation.

Direct study of dendritic properties has been attempted by a combination of intracellular recordings and intracellular staining; such a study has been carried out in the alligator cerebellum, in which the stain has shown an intradendritic location for some of the recordings (Llinás and Nicholson, 1971). A representative result is shown in Fig. 43 (left), in which responses are shown to parallel fiber volleys of increasing strength (A)-(C). The small spikes and bumps in these recordings were interpreted as all-or-nothing spike-like activity within one dendritic tree. The Purkinje cell dendrites are considered to have patches, or "hot spots", of excitable membrane; with a weak synaptic input, the patches fire singly (A) or asynchronously (B); with strong input they fire synchronously and summate thereby to give a large intradendritic action potential (C). It is presumed that the patches occur at branch points in the dendritic tree (Lorente de Nó and Condouris, 1949). As shown diagrammatically in Fig. 43 (center), the action potentials at branch points are presumed to spread electrotonically in a "pseudosaltatory" fashion, from branch point to branch point, toward the cell body, where they summate to produce an impulse that propagates down the axon.

This explanation for intradendritic impulse activity in Purkinje cells is essentially similar to that for spike-like activity in chromatolytic motoneurons (Chapter 5), for fast prepotentials in hippocampal pyramidal neurons (Chapter 11), and for small spikes in immature neocortical neurons (Chapter 12). Small potential components recorded from muscle spindles have also been interpreted as asynchronous impulse activity in individual nerve branches (Ottoson and Shepherd, 1972). In the case of Purkinje cells, as in the other cases, the explanation must be confirmed with a model that incorporates the particular geometry of the branches, the electrical parameters of the membrane, the sites of the postulated patches, and the kinetics of the active properties in those patches.

What, it may be asked, is the functional significance of active properties in the dendrites of Purkinje cells and other neurons? In addition to making distant synapses more effective vis-à-vis the impulse output from the axon hillock, impulse conduction in dendritic trees probably depends upon the proper amounts and timing of excitation and inhibition at successive branch points within the tree. As Rall (1970b) has pointed out, "such multiple possibilities of success or failure, at many different points of bifurcation, could lead to elaborate sets of contingent probabilities which would provide a single neuron (if it has suitable input patterns over the dendritic branches) with a very large logical capacity." The Purkinje cell provides a particularly vivid illustration of these possibilities.

DENDRITIC INHIBITION   The inhibitory neurons onto the Purkinje cells are differentiated into two types, stellate and basket cells. The stellate cell inhibition is directed to the Purkinje cell dendritic tree and, thus, provides yet another example of inhibition that has a dendritic, rather than a somatic, location. In line with the comments in Chapter 4, and the discussion of similar situations in the motoneuron (Chapter 5), mitral cell (Chapter 6), thalamic relay neuron (Chapter 9), and prepyriform pyramidal neuron (Chapter 10), we may presume that this siting is dictated by the dynamic relationship between the inhibitory and excitatory inputs to the dendritic tree. Figure 43 shows a model of this concept. The diagram shows that the stellate cells may be able to effect "a selective inhibition of particular dendritic segments of a Purkinje cell" (Llinás and Nicholson, 1971). In the view of these workers, the large changes in local conductance that accompany dendritic inhibition should produce a "functional amputation" of particular dendritic branches.

In contrast to the stellate cells, the basket cells deliver their inhibition to the axon hillock region of the Purkinje cells. In this position, it is optimally placed to strangle, as it were, the Purkinje cell output. For the Purkinje cell to be amputated by one neuron and strangled by another makes it appear to keep rather rough company. Perhaps a more felicitous image is that of "inhibitory sculpturing" (Eccles, 1964), that by the stellate cells being primarily a spatial shaping of activity within the dendritic tree, that by the basket cell being more in the temporal domain of impulse firing. This is with respect to the individual Purkinje cell; with respect to the population of Purkinje cells, both types of inhibition contribute to the spatiotemporal patterning of a multineuronal functional ensemble.

DENDRITIC SPINES A prominent feature of Purkinje cells is the profuse investiture of the dendritic branchlets with spines. We have seen that the parallel fiber synapses are exclusively to these spines, whereas the climbing fiber synapses are exclusively to the trunks and branches. This difference in synaptic input sites has a profound implication; it shows that the spine is not simply a devise for increasing the surface area of dendritic membrane but, rather, is differentiated as a specific anatomical site for a specific synaptic connection and function. A similar conclusion has been reached for the spines of neocortical pyramidal neurons (Chapter 12).

The Purkinje cell dendritic spines are notable in being entirely postsynaptic in position; there are no synapses from them onto neighboring neurons. In this respect, they resemble the spines of cortical neurons (see Chapters 10 through 12) and differ from the spines of olfactory granule cells (Chapter 6) and thalamic intrinsic neurons (Chapter 9).

The fact that the Purkinje cell spines are only postsynaptic in position means that their input-output functions are simpler than is the case for the presynaptic spines of olfactory granule cells and thalamic intrinsic neurons. The only output from such a spine is by current flow through the spine stem. Llinás and Hillman (1969) have pointed out that there must be a very high longitudinal resistance through the stem, with the consequence that even a small amount of postsynaptic current flow would give rise to a large postsynaptic potential. Furthermore, this high resistance lessens the shunting effect of a synapse on activity spreading past in the dendritic trunk. It can thus be seen that in the cerebellum, as elsewhere, a spine functions as a semi-independent input-output unit, and the elaboration of spines, therefore, serves to increase the complexity of the contingent probabilities that were mentioned above.

Despite all the interest that has been focused on the cerebellum, and all the data obtained, one is still left with the question, as Thach (1972) has put it: What does the cerebellum do? The best answer, at present, seems to be that it helps initiate and maintain some types of movement and posture and that some of the Purkinje cells may have specific relations to specific movements and postures. One of the most cogent theories to explain how this happens has been put forward by Marr (1969), and its interest for present concerns lies in the fact that it is formulated at the level of synaptic organization and hinges very much on the properties of the dendritic spines. The central hypothesis in this theory is that the synapses of the parallel fiber onto the Purkinje cell dendritic spines are modifiable; it is proposed that such modification, in the form of increased efficacy, comes about if a climbing fiber input to the Purkinje cell occurs at the same time as the parallel fiber input to the spine. The context of a movement is conveyed to a Purkinje cell by the pattern of mossy fiber-parallel fiber inputs to the spines; the Purkinje cell "learns" the appropriateness of its output for a movement by being triggered by the climbing fiber input; that learned behavior is stored ("memorized") by means of the increased efficacy of the parallel fiber synapses. As a result of the increased efficacy, the parallel fiber inputs themselves can activate the Purkinje cell and provide for its participation in the appropriate movement contexts.

It is beyond our present concerns to pursue the theory in greater depth. What is relevant to note here is that it basically involves a conditioning paradigm, in which one kind of input is not effective unless it is "learned" in conjunction with another input. We will mention a similar hypothesis in relation to the inputs to the hippocampus (Chapter 11). The conditioning paradigm has, of course, a long history in concepts of brain function, and, as formulated at the level of synaptic organization, it has invariably required the postulate that there is some change in synaptic efficacy through use (see Hebb, 1949). We have already pointed out (Chapter 3) that this is an active field of neurophysiological investigation and that there are some suggestive leads, but few conclusive results. The fact that the Purkinje cell response to a parallel fiber volley includes a long-lasting inhibition, perhaps due to persistent transmitter effects, might indicate the possibility of long-lasting changes at the synapses onto the spines, but, beyond that, the appetite of the theory outruns the nutriment of the data. The much more labile synaptic actions in the hippocampus (Chapter 11) and neocortex (Chapter 12) seem more likely on the face of it to provide for learned behavior.

9

THALAMUS

The thalamus is one of the key relay centers of the brain. It is most conspicuously the gateway to the neocortex, and as such it has evolved in close relation to the cortex. It is, therefore, like the cortex, most highly developed in mammals and, especially, primates. Pathways from all the major sensory systems (from the muscles, deep tissue, and skin; from the eye, the ear, and the tongue), with the single exception of the olfactory system, send their fibers here, to terminate on cells that in turn relay these inputs to the neocortex. Each part of the thalamus, in turn, receives fibers from the area of cortex to which it projects. Because of these relationships, the thalamic nuclei are integral parts of *thalamocortical systems* as well as of their *ascending sensory systems*.

Each sensory pathway has its specific thalamic nucleus; Fig. 44 shows in schematic fashion the position of each. The lateral geniculate nucleus (LGN) provides the relay for the visual input coming from the retina. The medial geniculate nucleus (MGN) provides the relay for the auditory input that arrives through multisynaptic pathways originating in the cochlea of the inner ear. The ventrobasal complex (VBC) is the center for somatosensory input arising in the skin, deep tissues, and muscles of the body, and relayed thither through the lower brain stem (see Fig. 1). The diagram of Fig. 44 illustrates a point that is important with regard to synaptic organization, and that is the very different positions of the thalamic relays in these pathways; there appears to be little similarity in the stages of processing prior to

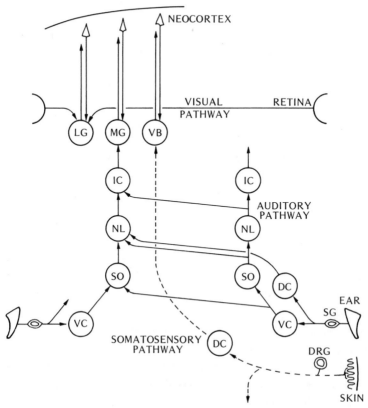

FIG. 44. Schematic diagram of different positions of the thalamic relay nuclei in their respective sensory pathways to the neocortex.

Visual pathway: lateral geniculate nucleus (LGN).

Auditory pathway: spiral ganglion (SG); dorsal cochlear nucleus (DCN); ventral cochlear nucleus (VCN); superior olive (SO); nucleus of the lateral lemniscus (NL); inferior colliculus (IF); medial geniculate nucleus (MGN).

Somatosensory pathway: dorsal root ganglion cell (DRG); dorsal column nucleus (DC); ventrobasal complex (VB).

the thalamic level in these three systems. The thalamus is, thus, a processing station for inputs from widely differing sources within the brain, a key fact for understanding its integrative position and for comparing it with other regions.

In addition to these relay nuclei in specific sensory pathways, there are other thalamic nuclei that provide relays to the cortex from the cerebellum and the hypothalamus. Still others receive inputs from various parts of the reticular formation and, in turn, project widely to the neocortex; they are often referred to as *nonspecific thalamic nuclei.* It may be noted that the term "nonspecific", used in this and other contexts, is largely a reflection of our ignorance of their connections and functions. As more is learned about them, the terms "diffuse", "multimodal", or "specific" may become applicable in many of these cases.

It should be evident from even these brief remarks that the thalamus is, in fact, a most complex group of relay centers. When we single out the specific sensory relay nuclei for study, as we shall now do, we must keep in mind that they constitute scarcely one-eighth of the total thalamic volume. It might appear that even the three sensory relay nuclei are too disparate to be discussed together in one chapter. In the face of this it is, therefore, all the more remarkable that in the studies of recent years, many workers have reported similarities in the synaptic connections and functional properties of the neurons in these three nuclei. It is, thus, precisely at the level of synaptic organization that evidence for common principles in the construction of these centers has been gained. In reviewing that evidence, we will focus on the LGN, and compare the findings there with those in the MGN and VBC.

NEURONAL ELEMENTS

In depicting the neuronal elements of the sensory relay nuclei of the thalamus, we first encounter the problem that the input (through the afferent fibers) arrives from "below" (caudally), while the output is directed to the cortex "above" (rostrally). This is the reverse of the diagrammatic convention we have sought to develop as a basis for comparisons between different regions. It is perhaps just as well to be reminded by this that the brain is too complicated to permit simple generalizations about the geometrical relations of its constituents. It will hopefully be least confusing to depict the thalamic neuronal elements in their true relations, as just described, and then follow our customary procedure when we come to the basic circuit, in order to facilitate comparisons there with the circuits of other regions. Our description of the neuronal elements in the LGN is based on the accounts of Guillery (1971), Famiglietti and Peters (1972), Rafols and Valverde (1973), and Grossman, Lieberman, and Webster (1973).

INPUTS The *afferent input* is through the specific sensory fibers. For the LGN, these are the axons of the optic nerve that arise from the ganglion cells of the retina, as described in Chapter 7. In most mammals, the axons that arise from the cells in the lateral parts of the retina pass to the LGN of the same side, while those from the medial part of the retina cross in the optic chiasm to the LGN of the other (contralateral) side; this arrangement is indicated only very schematically in Fig. 45. In the optic nerve the axons vary in diameter (1-10 μm); in higher vertebrates most or all of the axons are myelinated (see Chapter 7).

Within the LGN, the axons from the two sides terminate in separate layers; this is shown, again very schematically, in Fig. 45(A). Within its layer, the axon arborizes repeatedly to form many short branches, each of which ends in a knob-like terminal, as is shown in Fig. 45(B). The axonal branches are described as thick and myelinated up to very close to the terminal. The field of arborization is relatively restricted, of the order of 100-200 μm, in diameter. This restricted field is assumed to be a necessary basis for preserving the high degree of spatial acuity in the visual system.

The other type of input to the thalamus is the *central input* from the neocortex. This comes through axons from cortical pyramidal cells, which are described in Chapter 12. The axons are termed *corticothalamic fibers*. As already noted, these fibers arise largely in the same areas of the cortex to which the sensory nuclei project. They may thus be regarded as centrifugal fibers feeding back to their input source. These axons are described as fine and thinly myelinated. Within the LGN, they branch more

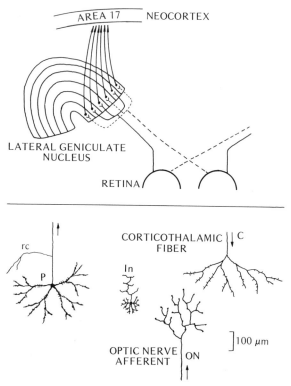

FIG. 45. *Above,* relation of layers in the geniculate nucleus to inputs from ipsilateral and contralateral retinas. *Below,* neuronal elements of the lateral geniculate nucleus.

Inputs: optic nerve afferent (ON); central fiber from cortex (C).

Principal neuron: thalamocortical relay neuron (P), with a recurrent axon collateral (rc).

Intrinsic neuron: (In) (shown with a short axon).

sparingly but more widely than the specific afferents. The branches have a distinctive appearance, in that they give off many short side branches during their course; see Fig. 45(B). These corticothalamic fibers are the main type of central input we will consider; other central inputs include fibers from *other thalamic nuclei* and, to varying extents, from such *brain-stem* regions as the reticular system.

PRINCIPAL NEURON   The ouput from the thalamic relay nuclei is carried in the axons of cells that, rare among principal neurons, have remained noneponymous; they are usually referred to as *relay*, or *thalamocortical, cells.* They take on a variety of forms in the different nuclei, but, in general, they appear, as in Fig. 45(B), as multipolar neurons, with four to eight dendrites that radiate outward in several directions. As principal neurons go, they are of small-to-medium size: the cell body varies in diameter from 15-30 μm, whereas the dendritic trunks vary from 1-8 μm in diameter. The dendrites are generally sparsely branched near the cell body, but they often have more branches peripherally. This gives them a tufted appearance (Ramón-Moliner, 1962), although this is not so marked as for the glomerular tufts of olfactory bulb neurons. Clusters of small appendages—knobs, thorns, spines—arise from the proximal and intermediate dendritic branches, as is shown in Fig. 45(B). The tree attains extents of 100-300 μm from the cell body.

The axon of the principal neuron may, or may not, give off collateral branches; this is still a matter of controversy (see Grossman et al., 1973). As we will see, such collaterals have figured prominently in concepts of functional organization of the thalamus, but it appears from the anatomical evidence that collaterals are not in general highly developed (cf. Schiebel and Schiebel, 1970), and in the monkey LGN they have been reported to be absent (Le Vay, 1971). In view of this fact, it seems particularly significant that the retinal ganglion cells, which provide the input to the LGN, are also conspicuous for their lack of axon collaterals (Chapter 7). This sequence of principal neurons in the visual pathway therefore provides a contrast to the well-developed systems of such collaterals in the output neurons of many other brain regions (e.g., mitral cells, motoneurons, Purkinje cells, cortical pyramidal cells).

The thalamocortical cell axons immediately enter the white matter of the cerebrum and travel to their projection sites in the neocortex. Within the white matter, they mingle with fibers of many other systems and, therefore, have not been accurately

characterized as to numbers and diameters. Most if not all of
them are presumed to be myelinated.

INTRINSIC NEURONS   A variety of small neurons has been described within the LGN of different mammals (Tömböl, 1967;
Guillery, 1971; Famiglietti and Peters, 1972; Rafols and Valverde,
1973; Grossman et al., 1973). In general, they are all subsumed
under the term *short-axon cell*, or *Golgi type II neuron*. Some of
these are very small (cell body 6-8 μm in diameter); an example
is shown in Fig. 45(B). The dendritic trunks are relatively thin
(1-3 μm in diameter), and the dendritic fields are limited to diameters of 50-150 μm. The dendrites are notable for giving rise
to clumpy, claw-like appendages along their course and at their
terminals. Other intrinsic neurons are larger, with cell bodies 10-
20 μm in diameter, and dendrites that are correspondingly
thicker and longer. In some species, the intrinsic neurons are described as bipolar in form, with their dendritic trees spanning
the layer in which they lie (Rafols and Valverde, 1973).

The axons of the intrinsic neurons are thin and unmyelinated.
They branch and terminate after a relatively short trajectory,
which is rarely more than 200 μm. These axons are among the
shortest in any region we will study; taken together with the cell
body size, they are then among the smallest neurons of any region. In some cases, the axonal field is in the vicinity of the cell
body and closely overlaps the field of its own dendritic tree.
Such a cell meets the classical definition of a short-axon cell, as
one whose axon terminates in the vicinity of its cell body. In
other cases, the axonal and dendritic fields do not overlap; see
Fig. 45(B); these fall under the category of short-axon cell as re-
defined in Chapter 6.

It has been reported that in monkey LGN the intrinsic neuron
lacks an axon altogether (Le Vay, 1971), which would put it in
the category of an *anaxonal* cell together with the retinal amacrine cell and the olfactory granule cell. This is a significant
finding, particularly in relation to the evidence regarding the synaptic connections of thalamic intrinsic neurons, as we shall pres-

ently see. As negative evidence, it awaits further study and confirmation.

The neuronal elements in the other thalamic nuclei are equivalent to those in the LGN, although there are many different patterns of dendritic and axonal arborizations. The reader may consult Morest (1971) for description of the MGN nucleus and Ralston and Herman (1969) and Schiebel and Schiebel (1970) for the VBC. Of particular interest is the fact that, just as in the LGN, axon collaterals of principal neurons, and axons of intrinsic neurons, are difficult to visualize. Even taking account of the difficulties of Golgi impregnation of thin unmyelinated axons, it appears that these constituents may be absent in some nuclei of some species (see Schiebel, Davies, and Scheibel, 1972).

CELL POPULATIONS   The number of optic nerve fibers to the LGN in man is approximately 1,000,000 (see Chapter 7). The total number of neurons in the LGN has been estimated to be approximately 1,000,000. The majority of these are principal neurons (see below), so it is permissible to conclude that the convergence ratio of afferents onto principal neurons is not much more than one. This approaches the strict 1:1 ratio of climbing fibers to Purkinje cells in the cerebellum, but stands in contrast to the much larger convergence ratios in most other regions of the brain (for example, retina, olfactory bulb, granule layer of cerebellum). Since the afferent fiber arborizations are relatively restricted, the input-output relations in the LGN are relatively point-to-point in the spatial domain. As already noted, this is assumed to be essential for the transmission of precise spatiotemporal patterns relayed thereto from the retina.

Some figures for relative numbers of principal and intrinsic neurons are available from recent Golgi studies. It has been reported that the ratio of principal:intrinsic neurons is approximately 10:1 in LGN (Le Vay, 1971), 2:1 in MGN (Morest, 1971), and 4:1 in VBC (Ralston and Herman, 1969). The remarkable feature of these ratios is, of course, that they reflect relatively small populations of intrinsic neurons in all the sensory relay nuclei. In this respect, there is a similarity to the cerebellum and probably also the ventral horn, and a contrast with the retina and olfactory bulb. The relatively small sizes and extents of the intrinsic thalamic neurons further restrict the possibilities for convergent and divergent arrangements vis-à-vis the principal neurons. These considerations, by themselves, indicate that the degree of processing carried out within the relay nuclei is likely to be rather modest.

SYNAPTIC CONNECTIONS

Electron-microscopic study of the synaptic connections in the thalamus has developed in three stages. First was the discovery that the main sites of synaptic connection are in tight clusters of terminals, called variously *synaptic nests, islands,* or *glomeruli.* Their general similarity to the synaptic glomeruli of the cerebellum was early noted (Szentágothai, 1963). The second stage involved the identification of the terminals within the glomerulus according to the classical criteria for axon and dendrite (see Chapter 1). This work established the basic similarity of the organization of the glomeruli in the three relay nuclei (Peters and Palay, 1966; Jones and Powell, 1969; Szentágothai, 1970). In the final, and very recent, stage, the relevance of the findings in olfactory bulb and retina has been realized, and certain of the presynaptic terminals have been re-identified as dendritic rather than axonal (Ralston and Herman, 1969; Morest, 1971; Famiglietti and Peters, 1972; Pasik, Pasik, Hámori, and Szentágothai, 1973; see also Reese and Shepherd, 1972). For our description, we will continue to focus on the example of the LGN.

A synaptic glomerulus in the LGN is illustrated in Fig. 46. At or near the center is the large terminal of an optic nerve axon. Around this terminal are numbers of other terminals; small glomeruli have only a few terminals; the largest have 15 to 20. The whole group or cluster is surrounded by one or more layers of glial membrane. Note that the size of a thalamic glomerulus is much smaller than a cerebellar glomerulus (Chapter 8, Fig. 40),

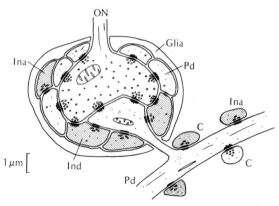

FIG. 46. Synaptic connections in thalamic sensory relay nuclei (lateral geniculate). Diagram shows a synaptic glomerulus, which contains axodendritic connections from an optic nerve terminal (ON) onto principal cell dendrites (Pd) and intrinsic cell dendrites (Ind), and dendrodendritic connections from intrinsic cell dendrites onto principal cell dendrites. Note also the axodendritic connections from intrinsic cell axons (Ina) onto principal cell dendrites. The glomerulus is surrounded by glial membranes. Outside the glomerulus are axodendritic connections from corticothalamic (C) and intrinsic cell axons onto a principal cell dendrite. (After Szentágothai, 1970; Jones and Powell, 1969; Morest, 1971; Famiglietti and Peters, 1972.)

and that both are microglomeruli in contrast to the macroglomeruli of the olfactory bulb (Chapter 6) and cerebral cortex (Chapter 12).

Within the glomerulus, the optic nerve terminal makes synaptic connections onto two types of dendritic terminal. One of these is a thorn or other appendage from the dendrite of a principal neuron; this is a type I chemical synapse (see Fig. 46). The other, and more numerous, type is constituted by terminal "knobs" or other appendages from the dendrites of intrinsic neurons; these connections are also type I chemical synapses. In addition, these terminals have dendrodendritic synapses, of type II, onto the dendritic thorns of the principal neurons. These connections are all indicated in Fig. 46.

When these arrangements were first worked out by electron

microscopists, the terminals that make type II synapses were assumed to be axon terminals of the intrinsic neurons, on the basis that they were presynaptic in position and, therefore, according to the functional concepts of the classical neuron doctrine, must, perforce, be axonal. As discussed in Chapter 2, these concepts are no longer tenable. With the demonstration in the olfactory bulb (Chapter 6) and retina (Chapter 7) that dendrites could occupy presynaptic positions, the identification of type II terminals in the thalamus as arising from intrinsic dendrites has become accepted. The intrinsic neurons thus receive axodendritic synapses from the optic nerve terminals and, themselves, make dendrodendritic synapses onto the principal neuron dendrites that also receive axodendritic synapses from the same optic nerve terminal.

In addition to these main patterns of connection, reciprocal dendrodendritic synapses have also been reported (Famiglietti, 1970; Harding, 1971; Lieberman & Webster, 1973). A few terminals from the axons of the intrinsic neurons are situated toward the periphery of the glomerulus and make type II synapses onto either type of dendritic terminal (principal or intrinsic). The possibility of axoaxonal synapses within the glomerulus has not been ruled out, but in the recent studies it is assumed that they are relatively infrequent.

Outside the glomeruli, two main types of connection have been identified. The more numerous are the connections from the terminals of extrinsic axons from the cerebral cortex (corticothalamic fibers) onto the dendrites of both principal and intrinsic neurons. They are type I synapses. These terminals can be identified unequivocally because they can be seen to degenerate after ablations of the cortex. There are also type II synapses onto the principal cell dendrites from the axons of intrinsic neurons. These connections are indicated in Fig. 46.

The patterns of connections in the other relay nuclei bear a general similarity to those illustrated in Fig. 46. In the MGN, the central element is usually the dendritic terminal from the principal neuron, rather than the axonal terminal from the afferent fiber; but, otherwise, the synaptic glomeruli are reported to have the same basic construction (Jones and Powell, 1969; Morest, 1971). The same can be said of the VBC, where the glomeruli have been reported to be somewhat less clearly demarcated by enveloping glial processes than in the other nuclei (Ralston and Herman, 1969).

BASIC CIRCUIT

The organization of the sensory relay nuclei of the thalamus is summarized in the basic circuit diagram of Fig. 47. In brief, the large terminals of the afferent axons make synapses onto the dendrites of both principal and intrinsic neurons within the synaptic glomeruli. The dendrites are interconnected by dendrodendritic synapses, predominantly oriented from intrinsic to principal neurons. The intrinsic neuron also has synaptic connections through its axon (when present) onto the dendrites of neighboring principal neurons. Depending on the presence of axon collaterals, there may be additional connections from the principal neuron onto the intrinsic neurons (Fig. 47, dotted line). These various connections provide for feedforward and feedback pathways within the thalamus. In addition, corticothalamic fibers have synapses onto the dendrites of both principal and intrinsic neurons, providing for long feedback loops through the cortex.

Much of the basic function of the thalamus in relaying information can be seen to revolve about the synaptic glomeruli, and it is appropriate, therefore, at this point, to assess their significance in the light of what we have learned about similar arrangements in other regions of the brain. Of particular interest in this respect are the cerebellar glomeruli, and Table 3 lists some of the salient features of these two cases. It may be immediately recognized that the thalamic glomeruli resemble their counterparts in the cerebellum not only in being surrounded by glial membranes, but, more importantly, in providing for specific interconnections between input, principal, and intrinsic elements. This relation was recognized by Famiglietti and Peters (1972) and designated the "triadic relation," and this is precisely the sense in which we

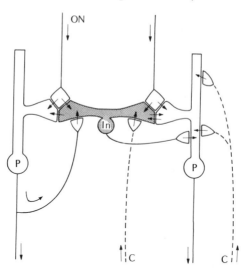

FIG. 47. Basic circuit diagram for thalamic sensory relay nuclei. Abbreviations as in Fig. 46.

have used the term in our discussion of the synaptic organization of the cerebellum and of other brain regions.

In view of this general similarity, it is of interest to compare, in greater detail, the organization of the thalamic synaptic glomeruli with those in the cerebellum. Table 3 lists some of the salient features of these two cases. It is obvious from a comparison of the lists that the two types of glomeruli have many features in common.

The shared features listed in Table 3 suggest that the granule layer of the cerebellum and the relay nuclei of the thalamus may be similarly placed as staging or processing stations for relaying inputs from different sources to a cortical site. A significant aspect of this comparison is that, whereas the granule cell population is everywhere homogeneous (as far as is known), in the thalamus, the neural substrate is differentiated into distinct regions, with specific variations on the common pattern of synaptic organization; this is, presumably, related to the specific functional demands of the relays in the three sensory pathways.

Table 3

Comparison of synaptic organization of granule cell layer of the cerebellum and sensory relay nucleus of the thalamus. The cerebellar granule cell is considered a principal neuron, by virtue of its relay function, and the cerebellar Golgi cell is considered an intrinsic neuron (see Chapter 8)

CEREBELLUM	THALAMUS
1. Large AFF (mossy) terminal	1. Large AFF terminal
2. Triad of AFF-PN-IN	2. Triad of AFF-PN-IN
3. AFF input to both PN and IN	3. AFF input to both PN and IN
4. Input-output divergence (1:10)	4. Input-output divergence (?)
5. DD desmosomes (PN-PN)	5. DD synapses (IN→PN)
6. IN has type II axon terminal onto PN	6. IN has type II axon and dendritic terminals onto PN
7. IN inhibitory to PN	7. IN inhibitory to PN
8. AFF's heterogeneous	8. AFF's heterogeneous (but homogeneous for one nucleus)
9. PN population homogeneous	9. PN population homogeneous (?)
10. One output site (molecular layer of cerebellum)	10. One output site (cerebral cortex)
11. No PN axon collaterals	11. Few or no PN axon collaterals
12. Little feedback from output site in cerebellum	12. Much feedback from output site in cortex

Key: PN, principal neuron; IN, intrinsic neuron; AFF, afferent input; DD, dendrodendritic.

The cerebellar and thalamic glomeruli provide a useful paradigm of the way in which knowledge at the synaptic level can provide the basis for studying principles of organization in different parts of the brain. This comparative approach must not be abused; its heuristic value lies not in the conclusion that different regions are similar, but rather in the insight it can give into the significance of the specific differences between the regions, as seen against the background of the features held in common. In the case of the cerebellar and thalamic glomeruli, some of the differences are quantitative: the number of afferent terminals, the number of principal and intrinsic neuron terminals, and the amount of divergence from the former to the latter, for example. Other differences are qualitative, and, of these, the presence of dendrodendritic synapses in the thalamic glomeruli appears to be among the most significant. It seems reasonable to infer that these provide for more complex patterns of interaction, and that more complex operations are, therefore, carried out in relaying information through the thalamic nuclei than through the granule layer of the cerebellum. This conclusion is also supported by the presence of recurrent axon collaterals of the principal neurons in at least some thalamic nuclei (the cerebellar granule cells have none), and the numerous extrinsic feedback connections through thalamocortical fibers onto principal and intrinsic neurons in these nuclei.

In comparison with other regions of the brain, the organization of the thalamus shares certain features with both the OPL and IPL of the retina, principally in that the input is made simultaneously from an input terminal to both principal and intrinsic elements. At both these levels in the retina, there are interconnections between the principal and intrinsic elements as well as back onto the input terminal; these interconnections make the arrangement of the synaptic triad more complex than in the thalamic nuclei. A thoughtful comparison between the synaptic organization of intrinsic neurons in the retina and thalamus has been provided by Morest (1971). A further step in complexity is seen in the macroglomerulus of the olfactory bulb, in which the inputs to principal and intrinsic elements are from separate afferent terminals, and reciprocal as well as serial dendrodendritic connections are present.

Apart from the synaptic glomeruli, the other prominent aspect of thalamic organization is the input from the cortex. The difference between afferent input, on the one hand, and central "centrifugal" input, on the other, is very clear and unambiguous. Not only are these fiber constituents separate (as they were not in the cerebellum), but their synaptic terminations are also distinct. It appears that the corticothalamic connections are primarily concerned with modulating the thalamic principal neurons through their dendritic branches, rather than controlling the initial events in the relay through the synaptic glomeruli. Like all other regions of our present study, except the retina, the thalamus has a feedback from the sites of projection that is integrated within its local circuits.

It is, of course, possible that, under some conditions, the cortex becomes the dominant input to the thalamic relay neurons, and the afferent fibers function as long and complex delayed feedback loops from the periphery. We must, therefore, be cautious about adopting an inflexible view of which inputs provide the primary drive to a region and which inputs provide feedback modulation; the same input pathway may function in either mode under different conditions.

SYNAPTIC ACTIONS

Since the thalamic relay nuclei receive specific sensory afferents, synaptic actions can be investigated by using brief sensory stimuli, as, for example, a spot of light, a click or tone, or a brief mechanical deformation. Alternatively, an electrical shock can be delivered to the afferent pathways. By either method, a synchronous volley is set up, similar to the method of procedure in other regions of the brain. We cannot review the many studies that are relevant to the evidence for synaptic actions in all the nuclei or even in one of them. Our concern must be limited to only the basic types of synaptic actions.

The key finding is illustrated in Fig. 48(A). An afferent volley produces an initial depolarization of the principal neuron. This depolarization has a brief latency, is graded in intensity, and, at threshold, gives rise to a single (or sometimes several) impulse(s) (see below); these are the characteristics of a monosynaptic EPSP. It is followed by a long-lasting hyperpolarization, during which impulse activity is completely suppressed; this suppression has the characteristics of an IPSP.

This type of excitatory-inhibitory sequence has been reported

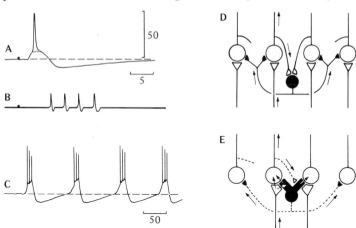

FIG. 48. Synaptic actions in thalamic sensory relay nuclei. (A) Intra-cellular recording of the response of a principal neuron to an afferent volley. (After Nelson and Erulkar, 1964; Andersen et al., 1964.) (B) Extracellular recording of the response of a presumed intrinsic neuron to an afferent volley. (After Andersen et al., 1964.) (C) Rhythmic "burst" responses of a principal neuron. (After Purpura and Cohen, 1962; Andersen and Eccles, 1962.) Intracellular voltage calibration, as in (A). (D) Schematic diagram of pathways for the generation of rhythmic activity through axon collaterals and short-axons. (Modified from Andersen and Eccles, 1962; Andersen and Andersson, 1968.) (E) Schematic diagram of possible pathways for rhythmic activity, in the light of recent evidence (see text).

in all relay nuclei. For references for the LGN, the reader may consult Burke and Sefton (1966), McIlwain and Creutzfeldt (1967), and Kalil and Chase (1970); for the MGN; Nelson and Erulkar (1964) and Aitken and Dunlop (1969); for the VBC, Andersen, Eccles, and Sears (1964). It is particularly notable that later authors have stressed the similarity of the responses in the three nuclei (see also Purpura, 1967, 1972; Eccles, 1969).

INHIBITORY PATHWAYS   The pathway that mediates the inhibition of the principal neuron has been investigated by recording from units believed to be the intrinsic neurons. These units respond to an efferent volley at a latency that is slightly later than the spike, but earlier than the IPSP, in the principal neuron; see Fig. 48(B). It has therefore been inferred that an impulse invades the axon collaterals of the principal neurons and synaptically excites the intrinsic cells, which then synaptically inhibit the principal neurons through their short axons (see Eccles, 1969). Such a pathway is analogous to the Renshaw pathway of the ventral horn, and similar pathways have been inferred in other regions, for example, the cerebellum and the hippocampus.

These physiological studies have not taken into account the recent findings regarding synaptic connections. As reviewed in the previous section, it is obvious that the dendrodendritic synapses also provide pathways for inhibition of the principal neurons. When we take into account the paucity of axon collaterals in most principal neurons and the equivocal evidence in some cases for short axons, the dendrodendritic connections within the synaptic glomeruli would appear likely to play a major role in the inhibitory control of principal neurons, particularly in their responses to afferent inputs (Morest, 1971; Shepherd, 1972b; Pasik et al., 1973). The role of afferent (lateral) inhibition as the basis for contrast enhancement in the specific sensory pathways has long been recognized (see Mountcastle, 1966). Physiologists now need to incorporate the dendrodendritic pathway into concepts of these processing mechanisms.

A similar re-evaluation is needed of the evidence for presynaptic inhibition. It was postulated that this was mediated through axoaxonal contacts from the intrinsic neurons onto the afferent terminals (Andersen et al., 1964), but the anatomical evidence suggests, instead, that the most likely pathway is through dendrodendritic inhibitory synapses from the inhibitory interneurons onto the principal neurons within the synaptic glomeruli (see Fig. 46).

Inhibitory potentials are also elicited in principal neurons by a volley from the area of cortex to which the neuron projects. Such potentials can still be recorded after cortical ablations, which cause the fibers from the cortex to degenerate. This was taken as evidence that this inhibition is also due to the Renshaw circuit within the thalamus (see Andersen et al., 1964). There is now the additional possibility that the antidromic impulse invades the dendrites of the principal neuron to activate excitatory dendrodendritic synapses onto the dendrites of the intrinsic neurons, which then feed back inhibition onto the principal neuron dendrites, in analogy with the sequence of mitral-granule-mitral inhibition in the olfactory bulb (see Chapter 6). But reciprocal synapses, and synapses from principal to intrinsic cell dendrites, are much less numerous in the thalamus than in the olfactory bulb.

From the experiments just cited, it has been further concluded that the inhibition caused by stimulation of the intact cortex is due to intrathalamic circuits and mechanisms (see Eccles, 1969). Recent work on the influence of the visual cortex on the LGN, however, has provided evidence for both excitatory and inhibitory effects (Kalil and Chase, 1970). The synaptic connections proposed for these effects consist of direct excitatory input to the principal neurons, and direct excitatory input to the intrinsic neurons, which are then inhibitory to the principal neurons. These connections are consistent with the basic circuit diagram of Fig. 47, with the exception that the latter connection is to presynaptic dendrites, rather than axons as was formerly believed.

RHYTHMIC ACTIVITY A prominent characteristic of thalamic neurons is the tendency to rhythmic activity. This is particularly evident under conditions in which neuronal excitability is high. As shown in Fig. 48(C), an afferent volley elicits, under these conditions, a burst of impulses in a principal neuron, followed by the long-lasting IPSP already described. The IPSP subsides over a period of 100 msec or so; when the membrane potential again reaches resting level, there is a rebound excitation, a burst of impulses is again generated, and the cycle repeats itself.

It was proposed that the rhythmically generated IPSP's are responsible for the phasing of the burst discharges (Purpura and Cohen, 1962) and that the IPSP's are due to feedback through a Renshaw type circuit (Andersen and Eccles, 1962). As illustrated in Fig. 48(D), the essential feature of the feedback circuit is that inhibitory interneurons have connections with many principal neurons, so that a discharge in a principal neuron leads not only to feedback inhibition of that neuron but also to feedforward inhibition of neighboring neurons. By this means, a large population of neurons becomes involved in synchronous excitation and inhibition within a few rhythmic cycles.

The mechanism as just described has been termed an *inhibitory phasing of neuronal discharge* by Andersen and Eccles (1962). It has served as a model for the generation of the alpha rhythm of the electroencephalogram (EEG) (see Purpura, 1967; Andersen and Andersson, 1968) and for the generation of rhythmic potentials by inhibition in other regions, for example the hippocampus (see Chapter 11). That it depends on pathways within the thalamus has been inferred from the fact that the rhythms persist after cortical ablation, which has been taken as evidence against the possibility of reverberatory thalamocortical circuits. That it depends on an intrathalamic pathway through axon collaterals and intrinsic cell axons has been postulated on the basis of an analogy with Renshaw inhibition in the spinal cord. But we have described in the olfactory bulb a possible mechanism for inhibitory phasing of neuronal discharge through exclusively dendrodendritic interactions between mitral and granule cell dendrites (see Chapter 6). The demonstration of dendrodendritic connections in the thalamus, as reviewed above, thereby provides alternative, or additional, pathways for rhythmical activity in the thalamus that now require physiological investigation. A tentative scheme to indicate how the dendrodendritic connections may be incorporated into the postulated circuits of the thalamus is shown in Fig. 48(E).

DENDRITIC PROPERTIES

PRINCIPAL NEURON   The evidence thus far indicates that the principal neurons of the thalamic relay nuclei are primarily postsynaptic in position. This means that in these neurons, as in other such neurons we have considered elsewhere, the dendrites provide for local integration of synaptic inputs, and transmission of the integrated potentials to the axon hillock. An electrotonic model is needed for assessing the properties of the dendritic tree in relation to these functions, but this is not available. The thalamic relay neurons are rather small, so that the acquisition of information about electrical parameters, upon which a model could be used, will not be an easy task.

There are a few clues to dendritic properties meanwhile. One clue is that, apart from details of geometry, the dendritic tree of the thalamic principal neuron is not unlike that of a smaller version of a motoneuron (Chapter 5) or of the glomerular tuft of an olfactory mitral cell (Chapter 6). By again making use of the concept of a scaling principle, it may be postulated that an equivalent cylinder for the dendritic trees of thalamic principal neurons would have an electrotonic length (L) in the same range of 1 to 2. We have seen that this is consistent with the transfer of signals through a dendritic tree by passive means alone, and it could be suggested, therefore, that this is also the case in the thalamic neurons. There is, in any event, little evidence for active properties of the dendrites, in terms of fast prepotentials or the like, as in cerebellar Purkinje cells or hippocampal pyramidal neurons. It is relevant to note, however, that following a period of inhibition the principal neurons become hyperexcitable, a phenomenon that is termed *post-inhibitory exaltation;* this is a possible additional component in the mechanism for the generation of rhythmic activity (cf. Purpura, 1967; Andersen and Andersson, 1968).

There is an interesting difference when the dendritic tree of the principal neuron is compared with the glomerular tuft of the olfactory mitral cell. In the thalamus, the afferent inputs are to the appendages on the more proximal dendrites, whereas the central (cortical) inputs are to the more distal branches. In the olfactory glomerulus, on the other hand, the afferent inputs are to the more distal branches and the interglomerular and centrifugal inputs are to the proximal branches. The relative placements of afferent and central inputs are thus different in the two cases, yet both have as primary functions the relay of specific sensory information and modulation by central pathways. This may serve as a useful caution against correlating particular functions with particular positions of synaptic inputs in a dendritic tree.

An interesting point bearing on dendritic properties is raised by the fact that the thalamic principal neurons respond to an afferent volley with an excitatory-inhibitory sequence [see Fig. 48(A)] similar to the response of the principal neurons of many other regions to an afferent volley. This similarity has seemed all the more remarkable in the face of the quite different sizes and shapes of the dendritic trees in the different neurons. To account for this most peculiar paradox, Purpura (1967) has suggested the principle that the major determinants of response patterns are the nature and distribution of postsynaptic potentials, rather than the size, configuration, and orientation of the dendrites.

This may seem inconsistent with our emphasis on the importance of dendritic properties and dendritic branching patterns in assessing the input-output dynamics of a neuron or neuronal part. Is it also at variance with such a statement as that of Peters et al. (1970): "the form of the dendritic tree provides a topographic map of the world as seen by a particular cell"? Probably not. One of the important results of our inquiry into dendritic properties is that the dendritic trees of most neurons are within 1 to 2 space constants of the cell body where recordings are usually made. Thus, regardless of the size or configuration of a neuron and its dendrites, postsynaptic potentials spread rather effectively through the tree, a fact that is implicit in the notion of a scaling principle (see Chapter 5). The summated response to a synchronous volley of many afferent fibers does not reflect the different sites of input in the tree. Under natural conditions, however, the contributions of individual inputs, weighted and shaped by the electrotonic properties of the dendritic tree, may be assumed to stand out more clearly within the integrative fabric of the neuron. Yet even under these conditions, the point raised by Purpura has some relevance. The particular geometry of a dendritic tree is our starting point for assessing dendritic function, but our final goal is the dynamic properties of the interactions within the dendrites.

INTRINSIC NEURONS   We have seen that the dendrites of the intrinsic neuron are presynaptic as well as postsynaptic in position. The dendrodendritic synapses from the presynaptic dendrites are not an occasional finding; they are a major type of connection, and we have reviewed the suggestion that, in some species, they may be the only output from the intrinsic neuron, if an axon is not present. This means that, as in the olfactory bulb and retinal neurons with presynaptic dendrites, the dendrites have a local input-output role to play as well as providing for longer-range transmission of signals within the dendritic tree.

These roles have been discussed in explicit and lucid form by Ralston (1971). Figure 49 is an illustration of the model he has suggested for dendritic functions in the intrinsic neurons. The starting point is the fact, which we have noted, that the dendritic terminals are at a distance from the cell body and axon hillock. This makes it possible for a dendritic terminal to receive synaptic input without the postsynaptic potential being of sufficient amplitude to spread to the axon hillock to generate an impulse. Ralston noted the possibility that the local postsynaptic potential would be adequate to activate local dendrodendritic synapses onto neighboring dendrites. Each of the terminals or branches might be capable of releasing its transmitter, graded by the amount of local synaptic depolarization, regardless of the activity of other dendritic branches; thus, each would function as an "independent synaptic unit". In contrast, when an action potential was initiated in the axon hillock, it might invade the proximal dendritic branches, causing synchronous release of transmitter from those synapses, but it probably would not reach the finer branches. In this way, several levels of output activity are possible in the synaptic output from the dendritic tree of the intrinsic neuron.

This model must now be tested in physiological experiments. As it stands, it is entirely consistent with the evidence for dendritic spread of an impulse in the olfactory mitral cell, with the

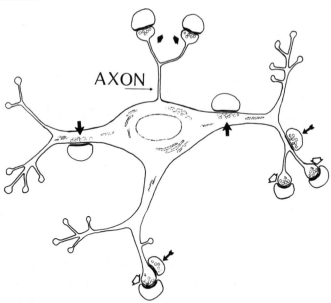

AXON

FIG. 49. Schematic diagram of an intrinsic neuron in the thalamus, to illustrate input-output relations in dendritic branches (large solid arrows) and dendritic terminals. Terminals can act as semi-independent functional units, receiving axonal input (solid, tailed arrows) and giving rise to local dendrodendritic synaptic output (open arrows). (From Ralston, 1971.)

evidence for graded transmitter release from retinal neurons and olfactory granule cells, and with the evidence for dendrodendritic synaptic interactions in the distal terminals of the periglomerular cell dendrites, as described in Chapter 6. The model of Fig. 49 may thus be taken, by and large, as a working model for the periglomerular cells in the olfactory bulb and for short-axon cells, with presynaptic dendrites, in other parts of the brain, as well as in the thalamus. The concept of independent synaptic units is further consistent with the suggestion that dendritic terminals form identifiable *functional units* at the level of synaptic organization (Shepherd, 1972b).

10

OLFACTORY CORTEX

We turn now to consider the synaptic organization of the cerebral cortex of vertebrates. From an evolutionary point of view, the olfactory cortex, defined as the part that receives the output fibers from the olfactory bulb, was the first to differentiate, becoming recognizable in the brains of primitive aquatic species. By virtue of this primal position, it is termed *palaeocortex;* later in the evolutionary sequence come the *archicortex* (hippocampus) and *neocortex* (the cerebral expansions of higher vertebrates). These three divisions will be the basis for our study. It is often stated that the olfactory cortex is the precursor or anlage from which the other types have been differentiated, but this is reading between the lines in the book of evolution, and the reader is referred to Herrick (1948) and Nauta and Karten (1970) for orientation to this question.

Of the three divisions, the olfactory cortex has been the least studied. One need not look far for the reasons. It is due, in part, to our persisting ignorance of the nature of the olfactory input. In part, it is the lesser role that olfaction plays in human behavior, compared with the auditory and visual systems. And, in part, it is a simple matter of the general inaccessibility of the olfactory structures, relegated as they have been to the antipodes of the brain by the overgrowth of the neocortex.

Despite its declining fortunes, the olfactory cortex is, in fact, the appropriate subject with which to begin the study of cortical

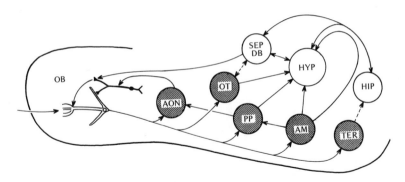

FIG. 50. Relations between olfactory and limbic parts of the brain. Primary olfactory areas, which receive input from mitral cells of the olfactory bulb (OB), are shown as shaded circles: anterior olfactory nucleus (AON); olfactory tubercle (OT); prepyriform cortex (PP); amygdaloid complex (AM); and transitional entorhinal cortex (TER). Limbic structures are shown as open circles: septum and diagonal band (SEP-DB); hypothalamus (HYP); and hippocampus (HIP). Note multiplicity of interconnections, including centrifugal pathways to granule cells of the ofactory bulb.

organization. Not only is it phylogenetically the primordial cortex, it is, by virtue of that fact, the principal cortical region of most lower vertebrate species. Indeed, the entire telencephalon of most lower vertebrates is dominated by the olfactory system. And it also plays this dominant role in many higher vertebrates. Evidence is accumulating that, despite the overgrowth of the neocortex, the sexual and social behavior of most mammalian species is primarily mediated by olfactory substances and pheromonal agents acting through the olfactory system (see Whitten and Bronson, 1969).

To these considerations is added the fact that the olfactory cortex is the simplest of the cortical regions in terms of its intrinsic structure. Having stated this, we must indicate more precisely which regions we have in mind when we refer to the olfactory cortex. The commonly accepted definition is in terms of the projection of the olfactory bulb (see Pribram and Kruger, 1954). Figure 50 shows this projection in schematic form; it can be seen

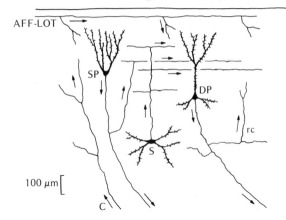

FIG. 51. Neuronal elements of the olfactory (prepyriform) cortex.

Inputs: mitral cell afferent axons in the lateral olfactory tract (AFF-LOT); fibers from central brain regions (C).

Principal neurons: superficial pyramidal neuron (SP); deep pyramidal neuron (DP); recurrent axon collateral (rc).

Intrinsic neuron: stellate cell (S).

that there are several sites that receive direct synaptic input from the bulb and qualify, thereby, as olfactory cortex. The largest site is the so-called *prepyriform cortex*, which is that region lying beneath and to either side of the lateral olfactory tract. This is usually regarded as the primary cortex in the olfactory pathway, and it will be the subject for our study.

NEURONAL ELEMENTS

The neuronal elements of the prepyriform cortex are shown in Fig. 51. We follow the descriptions of Calleja (1893), Cajal (1911; 1955), O'Leary (1937), Valverde (1965), and Stevens (1969).

INPUT   The *afferent* input is from the olfactory bulb, through the mitral cell axons in the lateral olfactory tract (LOT). The LOT lies as a band of fibers on the surface of the cortex, from which numerous branches are given off to the underlying and surrounding cortex. The branches fan out over the surface and reach considerable lengths (several millimeters). The branches all terminate in a superficial molecular layer of the cortex; none pass to the deeper layers. Thus, as in the olfactory bulb, the afferent fibers enter at the surface and terminate in a superficial layer. The situation is also similar to that in the hippocampus but differs from the cerebellum (Chapter 8) and neocortex (Chapter 12), in which all inputs arrive from the cortical depths. The axons in the LOT are thinly myelinated and range in diameter from about 0.5-3 μm (Allison, 1953).

Another type of input is through fibers entering the depths of the cortex; following the convention we have used elsewhere, it may be termed the *central* input. These fibers terminate at various levels in the cortex, except the most superficial (see Fig. 51). Their sources have not been identified. Association fibers from neighboring olfactory regions will be described later.

PRINCIPAL NEURON   The output from the prepyriform cortex is carried in axons that arise from several types of neuron. The major type is usually termed a *superficial pyramidal cell*, although, as Cajal (1955) noted, the cell bodies take on a variety of forms that may be semilunar, mitral, triangular, or polymorphic. The cell bodies are 15-30 μm in diameter, of modest size as principal neurons in the brain. As shown in Fig. 51, they are arranged in a distinct sheet some 400 μm below the cortical surface. Each cell has several dendritic trunks (several microns in diameter) that are directed superficially and arborize and terminate in the superficial molecular layer. The branches are moderately invested with spines.

These are usually termed *apical* dendrites. As such, they are regarded as analogous to the apical dendrites of neocortical pyramidal cells (see Chapter 12). However, a resemblance to the dendritic trees of granule cells in the dentate fascia (Chapter 11) has also been pointed out (O'Leary, 1937). The terms *pyramidal* and *apical*, like *granule* and many other terms we have encountered, must therefore not be taken to imply similar structures and functions when we compare neurons in different parts of the brain.

From some of these cells, one or more *basal* dendrites arise at the sides of the cell body; they branch sparingly and terminate within 100-200 μm of the cell body. They also are moderately invested with spines. Many of the superficial pyramidal neurons, however, do not have any clearly identifiable basal dendrites. This is noteworthy, since differentiation of the dendritic tree into apical and basal branches is a prominent feature of the pyramidal cells in the hippocampus and the neocortex. These prepyriform neurons are, therefore, distinctive in being cortical "pyramidal" neurons that lack basal dendrites.

The second, less numerous type of principal neuron is the *deep pyramidal neuron*. These neurons lie scattered in the zone just deep to the sheet of superficial pyramidal cells. The cell bodies are somewhat larger (20-40 μm in diameter). Each gives off a stout *apical dendrite* (3-6 μm in diameter) that ascends into the molecular layer and ramifies there. Each also gives rise to several *basal dendrites;* they have bushy arborizations that are often directed toward the depth of the cortex, giving them a brush-like appearance (see Fig. 51). Both apical and basal dendrites are liberally invested with spines. Many of these cells resemble the small pyramidal neurons of the neocortex. In both cases, it should be noted, the term *pyramidal* has no necessary significance beyond the geometrical form of the cell body imparted by the characteristic origins of the apical and basal dendritic trunks (see Price, 1973).

A third, infrequent type of principal neuron is described as *polymorphic* or fusiform; these neurons lie in the depths of the cortex (see Fig. 51). The cell bodies range in size from 15-40 μm in diameter. Each gives off several dendritic trunks, which have different orientations and which ramify and terminate within the cortical depths over distances of several hundred microns.

The fact that the principal neurons of the prepyriform cortex have these different forms and locations is a notable feature which is obviously significant for the synaptic organization of this region. It is probable, for example, from the above description, that the more superficial neurons are primarily related to the afferent input, whereas the deeper neurons are primarily related to the central input. For our purposes, we will consider the pyramidal neurons, both superficial and deep, as forming the main population of principal neurons.

The axons of the pyramidal cells give off several collaterals within the cortex. Some of these collaterals terminate within the deep layers, either immediately or after extending for variable distances in the lateral direction, whereas other collaterals recur to the molecular layer (see Fig. 51). The latter collaterals are analogous to association fibers of the hippocampus (Chapter 11) and neocortex (Chapter 12); we will have more to say about their significance later. Some of the pyramidal cell axons do not leave the cortex; these cells qualify thereby as intrinsic neurons (see below; see also Chapter 12). The axons that do leave the cortex join the great mass of deep white matter and distribute to many regions. These include the surrounding olfactory regions and parts of the hypothalamus and thalamus (Powell et al., 1965; Heimer, 1968; Price and Powell, 1971) (see Fig. 50). By these connections, the olfactory cortex is strategically placed to affect central brain structures that play crucial roles in many types of behavior (see Chapter 11).

INTRINSIC NEURONS Scattered throughout the cortex are cells whose axons do not leave the cortex; these cells qualify, therefore, as intrinsic neurons, by our general definition. The majority are polymorphic cells in the middle and deep layers. They are often referred to as *stellate cells* from the star-shaped form imparted to the cell body by the several dendritic trunks. The cell bodies are 10-20 μm in diameter, and the trunks are several microns in diameter. The latter arborize over fields of 100-300 μm and are lightly invested with spines. A thin axon (1-2 μm thick) is directed toward the depths, the surface, or laterally, as the case may be; branches and terminals are given off at all levels (except the most superficial). Some of the branches extend in the lateral direction for considerable distances (up to a millimeter or more). It has been reported (O'Leary, 1937) that some stellate axons terminate in baskets around the cell bodies of pyramidal cells, similar to the baskets in the hippocampus (Chapter 11).

Despite the often considerable length of their axons, these cells may be termed *short-axon cells*, by our previous definition, on the basis that the axon remains confined to the cortex within which it arises (see Chapter 6). But as we have already noted, the stellate cells are part of the population of polymorphic cells, and there is a gradation in lengths and destinations of axons within this population. Possibly, the axons that enter the white matter do not re-enter the prepyriform cortex, but evidence on this point is lacking. It is therefore premature to draw a hard and fast line between principal and intrinsic neuronal types in this region. We have already mentioned this problem in our discussion of the neurons of the spinal cord (Chapter 5) and we will encounter it again for the neuron types of the neocortex (Chapter 12). The fact that some pyramidal cell axons do not leave the cortex should also be recalled in this regard.

CELL POPULATIONS   Quantitative studies of prepyriform elements are almost completely lacking. Such studies are hampered by the fact that the output axons do not form a single tract that can be subjected to quantitative analysis. It would appear that there are more pyramidal cells in the prepyriform cortex than the 60,000 or so LOT axons that have been estimated (Allison, 1953), so that the input-output ratio for this region is probably rather low. The input side is raised considerably by the large number of branches of the LOT fibers, but the convergence ratio would still appear to be less than 10:1. Within the cortex, intrinsic elements (the stellate cells) are much less numerous than pyramidal cells, so the ratio I:P must be less than 1:1, and might well be less than 1:10.

These very approximate figures are enough to suggest a low input-output convergence and low I:P ratio in the prepyriform cortex. This is similar to the case in the cerebellar cortex vis-à-vis the climbing fiber input (Chapter 8) and also the neocortex (Chapter 12). These regions stand in sharp contrast to the high convergence ratios and high I:P ratios in the olfactory bulb

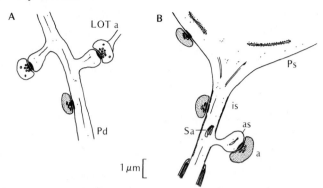

FIG. 52. Synaptic connections in the olfactory (prepyriform) cortex. (A) Axodendritic connections from LOT axons (LOT a) onto the spines of a pyramidal cell dendrite (Pd). (B) Axosomatic connections to the cell body of a pyramidal neuron; axoaxonic connections to the initial segment of an axon (is); and axoaxonic connection to an axonal spine (as) of the initial segment. Note also a spine apparatus (sa) in the initial segment. (After Westrum, 1970.)

(Chapter 6) and retina (Chapter 7). The ratios have significance for the kinds of signal processing carried out in these regions, as we will discuss below in connection with the basic circuit diagram for the prepyriform cortex.

SYNAPTIC CONNECTIONS

The main fact known about the synaptic connections in the prepyriform cortex is that the branches of the LOT fibers make exclusively axodendritic synapses onto the apical dendrites of the pryamidal cells. The synapses are made onto the dendritic shafts and spines in the superficial molecular layer; see Fig. 52(A). Most of the synapses have round vesicles in the presynaptic terminal and asymmetric membrane thickenings, i.e., they are type I chemical synapses (Lund and Westrum, 1966; Westrum, 1966a,b, 1970). The LOT axons arise from mitral cells in the olfactory bulb, and it will be recalled that, in the bulb, the mitral cells make type I chemical synapses at all levels: dendrodendritic in the olfactory glomeruli, dendrodendritic in the EPL onto granule cells, and axodendritic through the axon collaterals (see Chapter 6). The fact that the mitral cell also appears to make the same type of synapse at its afferent axon terminal in the prepyriform cortex has been adduced as evidence of the morphological corollary of Dale's Law (see Chapter 2) that a neuron secretes the same transmitter substance at all its synapses. A few terminals in the superficial layer are type II; it has not been determined whether they arise from extrinsic or intrinsic sources.

In deeper layers, type II chemical synapses predominate onto the cell bodies of the pyramidal neurons (see Fig. 52 B). Of particular interest is the finding of synapses onto the initial segments of the axons of these cells (Westrum, 1966b, 1970). These studies have also shown that there are spines arising from the initial segments, with a spine apparatus either within the spine or within the nearby initial segment. As shown in Fig. 52(B), the synapses onto both initial segments and spines are type II; classified in terms of the pre- and postsynaptic structures, they are axoaxonal. It seems likely that they arise, at least in part, from the basket arborizations observed in the Golgi-stained material.

These findings demonstrate several principles. They show that a spine, together with a characteristic spine apparatus, may arise from an axon as well as from a dendrite. They show that an axonal process—initial segment or spine—may receive synaptic inputs. We have discussed an instance of this in the cerebellum (Chapter 8), and similar synaptic arrangements have been observed in the neocortex (Chapter 12). These findings are, of course, inconsistent with traditional views of synaptic orientations and are additional evidence of the need to revise and enlarge our concepts of the functional organization of neurons.

Apart from these observations, little is known about synaptic connections in the prepyriform cortex; no dendrodendritic synapses have been described, and the sources of the connections have not been identified. The specific connections between neuronal types, therefore, have had to be inferred from the Golgi preparations and from the electrophysiological analysis of synaptic actions (see below).

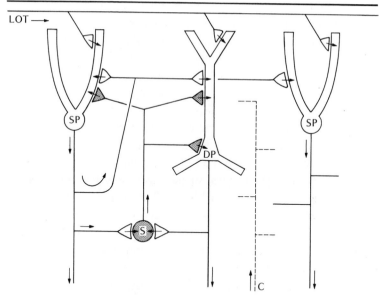

FIG. 53. Basic circuit diagram for the olfactory (prepyriform) cortex. Abbreviations as in Fig. 52.

BASIC CIRCUIT

The organization of the prepyriform cortex is summarized in the basic circuit diagram of Fig. 53, which is derived from the accounts of Biedenbach and Stevens (1969) and Haberley and Shepherd (1973). In addition to subsuming the anatomical evidence just reviewed, the diagram anticipates the physiological evidence for synaptic actions that is presented in the following section.

To summarize briefly, the branches of the LOT axons make synapses onto the apical dendrites of superficial pyramidal cells and, probably, the deep pyramidal cells. Both types of pyramidal cell, through their deep axon collaterals, make synapses onto stellate cells. It is also probable that through their long axon collaterals, either deep or recurrent, the pyramidal cells make synapses back onto the apical dendrites of other pyramidal cells. The axons of the stellate cells also distribute to the apical dendrites of the pyramidal cells, again over both short and long distances. Finally, little is known with regard to the central inputs coming through the cortical depths, but it is probable that these fibers make connections onto both the pyramidal cells and the stellate cells.

Let us now consider some aspects of synaptic organization within this basic circuit in more detail. To begin with, the connections within the prepyriform cortex are made within a strongly horizontal and vertical framework, as is emphasized by the diagram. The afferent input in the LOT propagates as a strictly horizontal wave across the cortical surface, as rigidly horizontal as the parallel fiber input to the Purkinje cell dendrites in the cerebellum. Moreover, this sequence is unidirectional for the entire cortex, always proceeding from anterior to posterior, away from the olfactory bulb. In the cerebellum, by comparison, each granule cell axon bifurcates to give rise to two parallel fibers that run in opposite directions. This means that, in the prepyriform cortex, there is a temporal sequence of activation by the afferent input, which is possibly even more rigid than the cerebellar sequence. A similar rigid sequence is present in the hippocampus (Chapter 11).

The afferent input to the prepyriform cortex is transferred synaptically to the ends of the apical dendrites of the principal neurons, similar to the way that the olfactory input is transferred to the terminal tufts of the mitral cells in the olfactory bulb. It seems highly significant that two successive relays of this same basic type (axodendritic, to distal dendritic terminals) should occur in succession. What the implication might be for information processing in the olfactory pathway, in contrast to other sensory pathways, is not immediately evident, but it can at least be concluded that vertical dispersion of inputs onto the principal neuron dendrites is one parameter of synaptic organization not used in this pathway.

It is further noteworthy that the afferent input through the LOT is transferred only to the principal neuron of the prepyriform cortex, the pyramidal cells. This may immediately be recognized as different from the close relation between the afferent, principal, and intrinsic elements of the synaptic triad we have seen in the olfactory bulb, retina, cerebellum, and thalamus. It is also different from the situation in the ventral horn and in the neocortex, in which the afferent fibers have direct access to intrinsic neurons. It is similar, however, to the case of the hippocampus. This would appear to be an important aspect of the olfactory relay in the prepyriform cortex. The central inputs, on the other hand, arriving through the cortical depths, have access to both the deep stellate cells and the pyramidal cells, so that the patterns of connections established over this route may involve the triad of synaptic elements. The two inputs—afferent and central—thus can make quite different use of the prepyriform neuronal circuits. It should be noted that this over-all picture is modified by the extent to which pyramidal neurons have axons that do not leave the cortex, and stellate cells have axons that do leave the cortex, as has been discussed previously.

The diagram of Fig. 53 indicates that there are two major types of intrinsic pathway within the prepyriform cortex. One consists of the circuit from axon collaterals of principal neurons through inhibitory intrinsic neurons back onto the principal neurons. There is an obvious similarity to the Renshaw circuit in the ventral horn (Chapter 5), in that this provides for inhibitory feedback of responses to afferent inputs. In relation to central inputs, on the other hand, the intrinsic neurons can provide a feedforward as well as feedback pathway. As is shown in Fig. 53, the deep position of most of the intrinsic neurons and their inaccessibility to direct LOT activation ensure their different use by these two inputs.

The other type of intrinsic pathway is by way of the long axon collaterals of the pyramidal cells. We have already suggested that because of their length they may be regarded as analogous to the association fibers of the hippocampus and neocortex. Recent studies have provided evidence that these collaterals make direct connections, of an excitatory nature, onto the pyramidal cells themselves. The evidence, both anatomical and physiological, will be discussed in later sections; here, it may simply be recalled that such *recurrent re-excitatory pathways*, mediated by axon collaterals of principal neurons onto other principal neurons, have been conspicuously absent in the regions of the brain thus far considered in this book. We will encounter this type of pathway in the hippocampus (Chapter 11), and in Chapter 12 we will discuss the evidence for such pathways in the neocortex. In the cerebellum (Chapter 8), the axon collaterals of Purkinje cells also connect to other Purkinje cells, but the connections are inhibitory. We note for now the possibility that re-excitatory circuits may be an important constituent of the organization of cerebral cortical regions.

SYNAPTIC ACTIONS

The LOT is a clearly defined bundle of axons that can be discretely stimulated with an electric shock. This, and the fact that the axons all make their synapses onto the distal dendrites of the pyramidal neurons, has permitted an electrophysiological analysis of synaptic actions over this route.

The single unit responses are illustrated in Fig. 54(A). Here an intracellular recording from a pyramidal neuron shows an initial depolarization lasting 10-20 msec; at threshold amplitude, it leads to the generation of an impulse. This is followed by a wave of hyperpolarization that cuts short the initial excitation and lasts for several hundred milliseconds if the volley is strong. It has been concluded that this sequence represents an initial EPSP followed by an IPSP (Biedenbach and Stevens, 1969; Haberly, 1973a). Observe that the sequence is very similar to that which we have seen in the principal neurons of other brain regions subjected to a single afferent volley. Compare it, for example, to the responses of thalamic neurons (Fig. 48); the similarity extends to the relative time courses of the excitatory and inhibitory periods.

Intracellular recordings from prepyriform neurons are difficult to obtain, and a comparison between the responses of the different neuronal types is therefore best carried out using extracellular unit recordings (Haberly, 1973a). In Fig. 54(B) is shown

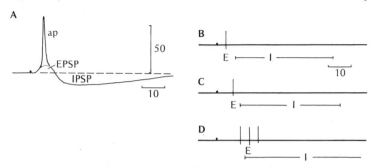

FIG. 54. Synaptic actions in the olfactory (prepyriform) cortex. (A) Intracellular recording of the response in a pyramidal neuron to a volley in the lateral olfactory tract. (After Beidenbach and Stevens, 1969; Haberly, 1973a.) (B) Extracellular unit recordings of the response of a superficial pyramidal cell to a LOT volley. Periods of excitation and inhibition, as revealed by a second test volley, are indicated by E and I, respectively. (C) Response of a deep pyramidal neuron. (D) Response of a deep stellate cell. [(B)-(D) after Haberly, 1973a.]

an extracellular recording of a superficial pyramidal neuron undergoing the same response as in (A). The spike rides on a large wave, which is the *evoked potential*, due to the summed extracellular currents of the entire population of active neurons. The horizontal bar indicates the time during which the response to a second testing volley is suppressed; note that this coincides with the time during which the cell is undergoing the IPSP, as shown in (A).

For comparison, the response of a deep pyramidal neuron is shown in Fig. 54(C). The synaptic action is similar, consisting of an initial excitation followed by a long period of unresponsiveness. The latency of the initial impulse is longer, however, allowing for a synaptic relay in the pathway of activation. The obvious candidate for this pathway is through the axon collaterals of the superficial pyramidal neurons. The physiological experiments have shown that there is indeed a wave of activity that spreads at a slow rate (0.8 m/sec) throughout the cortex after the initial LOT input, which is presumably due to impulses traveling through the long collaterals of the pyramidal axons. It has been concluded (Haberly, 1973b) that most deep pyramidal neurons respond to afferent input by this relay. A direct route cannot be ruled out for some deep pyramidal neurons; this connection is, therefore, shown by the dotted line in Fig. 53. The axon collaterals provide the pathways for re-excitatory circuits within the prepyriform cortex, as discussed later in relation to dendritic properties.

RECURRENT INHIBITION   The intrinsic neurons (stellate cells) also respond to an LOT volley. As shown in Fig. 54(D), the responses have relatively long latencies, indicating at least one relay through the pyramidal neurons. Also, these deeply situated cells tend to fire several impulses in response to a single volley. The excitatory phase is followed by a period of unresponsiveness to testing volleys, as shown in Fig. 54(D).

It has been concluded that the inhibition that is present in all three types of prepyriform neuron is mediated by the stellate cells acting as inhibitory neurons (Biedenbach and Stevens, 1969; Haberly, 1973a; Haberly and Shepherd, 1973). As indicated in the basic circuit diagram of Fig. 53, the stellate cell axons are directed laterally and superficially to make synapses onto both the superficial and deep pyramidal neurons. As already noted, the circuits thus formed are analogous to the Renshaw pathway in the ventral horn (Chapter 5). By inhibiting the pyramidal neurons, the stellate cells also cut off their excitatory input from the pyramidal neurons, as is shown by the unresponsiveness of the stellate neurons to a testing volley after their initial excitation [Fig. 54(D)]. The unresponsiveness is therefore not a direct IPSP onto the stellate cells themselves but rather an indirect action; it is, in fact, a variety of presynaptic inhibition of the excitatory pathway onto the stellate cells. This does not rule out the possibility that some stellate cells may also receive direct inhibition, through connections between each other.

RHYTHMIC ACTIVITY   The prepyriform cortex is notable for the prominent rhythmic potentials that can be recorded from it by electroencephalography, and the mechanism for the generation of these potentials was the subject of some of the earliest electrophysiological studies of this region (Freeman, 1959). The point of departure for these studies was the interesting fact that, with very weak shocks to the LOT, the threshold response of the prepyriform cortex takes the form of a low-amplitude oscillating field potential. Analysis showed that the response in the depths of the cortex was of similar form but opposite polarity to that recorded at the surface, indicating a potential dipole across the cortex. Using a systems engineering approach, Freeman (1962) developed a model for the oscillations, based on an excitatory input with intrinsic negative feedback. The negative feedback performed a gating function that gave rise to the rhythmic activity.

It may be recognized that this model is formally similar to that proposed for the generation of rhythmic potentials in the olfactory bulb (Chapter 6), thalamus (Chapter 9), and neocortex (Chapter 12). Indeed, stripped to its essentials, the model is applicable in some form or other to almost any rhythmic system, neural or non-neural. This qualifies its use as an analytical tool but, nonetheless, suggests that the prepyriform cortex shares certain properties of rhythmic potential generation with other regions of the brain.

Freeman (1968) further demonstrated that removal of the bulbar input eliminates the oscillations recorded from the prepyriform area; the effect appears to be due to the removal of a steady excitatory input from the olfactory bulb through the LOT axons. This is in accord with the model, which requires an excitatory background as the continous input that is then subjected to inhibitory gating. Unit recordings have shown that the pyramidal cells fire impulses in synchrony with the rhythmic waves. There is sometimes a double peak of excitation, ascribed by Freeman (1968) to a reactivation loop within the cortex. The experiments of Haberly (1973a,b) suggest that this may be the re-excitatory loop provided by the axon collaterals of the pyramidal neurons.

DENDRITIC PROPERTIES

Our interest naturally focuses on the apical dendrites of the pyramidal neurons. All the evidence to date indicates that these dendrites are exclusively postsynaptic in position. Although this does not rule out the possibility that dendrodendritic synapses may, in fact, be present, it does provide a starting point for assessing the properties of the apical dendrites in relation to local integration of synaptic inputs and transfer of signals to the axon hillock.

An electrotonic model is not available for the assessment of these functions, but the geometry of the pyramidal neurons is sufficiently stereotyped that the essence of a model can be sketched in outline. Such a model is depicted in Fig. 55. The upper diagram reproduces a Golgi stained pyramidal cell; note again the relatively small size of this cell as a type of principal neuron. Combining the apical dendrites yields an equivalent cylinder, as shown in the middle of the figure. There is no direct evidence concerning the electrotonic length of the cylinder, but if the electrical parameters for other neurons are used, a value of about one is obtained, as shown.

Let us consider first the simplest case of a synchronous synaptic input through the LOT. This input activates excitatory synapses onto the peripheral branches of the apical dendrites, giving rise to an EPSP. The electrotonic spread of this EPSP to the cell body and axon hillock, to trigger an action potential, which propagates into the axon, is depicted in the lower diagram in Fig. 55. Note that because of the relatively short electrotonic length of the apical dendrite, this spread is very effective by passive means alone. The reader will realize that the diagram gives the spatial distribution for the EPSP and the action potential at the instant of impulse generation shown in the recording of Fig. 54. There is no evidence for fast prepotentials (see Chapter 11) or other forms of active properties, in addition to the passive potentials depicted in Fig. 55.

Subsequent to this initial synaptic excitation is the recurrent inhibition through the stellate cells, as shown in Fig. 54. The elec-

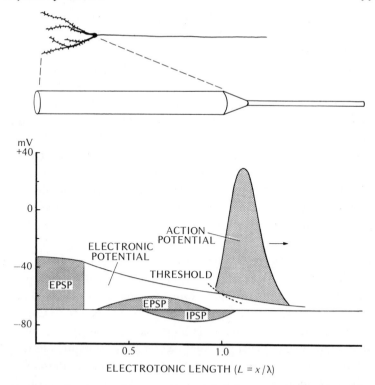

FIG. 55. Electrotonic model for a superficial pyramidal neuron of the olfactory (prepyriform) cortex, to show distribution of potentials for the case of an afferent input through LOT axons to superficial dendrites of the pyramidal neuron. Distribution of EPSP and IPSP due to intrinsic pathways is also shown.

trotonic model illustrates particularly vividly how this inhibition is not only located close to the cell body, to exert direct control over impulse generation there, but also is sited between the distal dendrites and the cell body, to most effectively gate, as it were, the transfer of the afferent EPSP to the axon hillock. This may be taken as yet another example of the strategic siting of inhibitory input relative to the site of excitatory synaptic input. Under some conditions, the nearness of the inhibitory site to the axon hillock may, in fact, not be sufficient for control over impulse initiation there; experiments have shown, for example, that inhibition cannot prevent an impulse from invading the cell body antidromically from the axon (Haberly, 1973a).

The second major excitatory input to the apical dendrites is through the axon collaterals of the pyramidal cells. This has come to light in the course of recent work, in which anatomical studies have demonstrated the presence of the collaterals (Heimer, 1968; Price, 1973) and physiological studies have independently demonstrated their synaptic actions (Haberly, 1973a,b; Haberly and Shepherd, 1973). The anatomical studies have shown that the collaterals from the pyramidal cells extend throughout the prepyriform cortex, terminating preferentially at the level of the proximal shafts of the apical dendrites, as has been shown in the basic circuit diagram of Fig. 53, and is indicated in Fig. 55. We have seen that this level is near one of the sites of inhibitory input from the stellate cells. Thus, the inhibitory input is sited in close proximity to the excitatory input from the long-axon collaterals. As in the case of the overlapping inputs to the motoneuron (Chapter 5), this provides for a direct clash between excitatory and inhibitory conductance changes that accompany synaptic activity.

Under more natural conditions of asynchronous afferent inputs, there must be a delicate balance between the intrinsic inhibition and the two types of excitatory input. Physiological recordings have, in fact, revealed both inhibitory and excitatory responses of prepyriform cells when the nose is stimulated with odors (Haberly, 1969). These results have suggested the possibility that there are sequences of processing in the olfactory cortex similar to those that have been identified in the sensory areas of the neocortex (see Chapter 12).

We may conclude by noting that some years ago Lorente de Nó (1938) stated that "the primary olfactory cortex is in fact a subcortical center comparable to the geniculate bodies, etc." This is a useful point to discuss, not only in view of the common belief that the olfactory areas are a part of the cerebral cortex, but also because it challenges us to define more precisely what is meant by the term *cortex*. By cortex, some authorities mean only the neocortex; some mean a region with a certain number of layers of cell bodies and nerve fibers; and some mean any part of the outside rind or covering layer of the brain. As with so many other terms, there are too many exceptions to any of these definitions; and, although the worker on the neocortex may be satisfied to use the term only as he defines it, this is no help to the student who is interested in comparing the organization of the neocortex with other parts of the nervous system.

In previous chapters (e.g., Chapter 6), it is been indicated that the study of synaptic organization may provide a basis for distinguishing between cortical and noncortical types of regions in the brain. Following this approach, we may note certain features of the prepyriform cortex that are distinctive in its synaptic organization: (1) there is a parallel orientation of apical dendritic trees of the principal neurons; (2) there is a non-repeating sequence of layers in relation to the vertical extent of the principal neurons; (3) the principal neurons are graded in terms of vertical extent and dendritic geometry; (4) the principal axon collaterals give rise to internal feedback circuits that are widespread and excitatory to principal as well as intrinsic neurons; and (5) inhibitory actions by intrinsic neurons into principal neurons are profound and long lasting. A columnar organization is a prominent characteristic of the neocortex (Chapter 12); whether there is a similar or analogous organization of prepyriform cortex is still a matter of conjecture (Stevens, 1969; Haberly, 1969).

These features are shared by the prepyriform cortex with other parts of the cerebral cortex—the hippocampus (Chapter 11) and the neocortex (Chapter 12)—and are absent, to a greater or lesser extent, in subcortical regions; we may cite as examples the spinal cord (Chapter 5) and the thalamus (Chapter 9). On the other hand, the prepyriform cortex shares with many subcortical regions the following features: (1) the principal neurons are small; (2) some principal neurons lack differentiation of the dendritic tree into apical and basal dendrites; and (3) the principal neurons receive afferent input that has not passed through the thalamus. It may be best to recognize, therefore, that the olfactory cortex combines features of both cortical and subcortical regions.

From these considerations, it is suggested that the study of synaptic organization provides a basis for distinguishing certain characteristics of local organization that may be termed cortical. That such a distinction may lead to the recognition that some regions with the anatomical label of nucleus may have an organization that has cortical characteristics, and that some cortical regions may have parts with a nuclear type of organization, is not inconsistent with this use of terms. What is significant is the extent to which otherwise diverse regions share such aspects of synaptic organization as the features mentioned above. It implies that certain modes of information processing are possible with a cortical type of organization that are not possible with the other, noncortical, type. This may be regarded as only a tentative proposal toward a rethinking of this problem, which now must be incorporated into the traditional views of the distinctions between cortical and nuclear regions.

11

HIPPOCAMPUS

The hippocampus is designated the archicortex, intermediate in evolutionary development between the olfactory palaeocortex and the neocortex. It is closely associated with a neighboring region, the *dentate fascia;* together they form an S-shaped structure (see Fig. 56), which reminded the early histologists of a sea horse (hippocampus) or a ram's horn (Ammon's horn).

In lower vertebrates, the hippocampus lies in a dorsal position in the brain, near the septal and hypothalamic regions. In the evolution of higher vertebrates, the hippocampus is dragged, as it were, in a long arc through the brain, coming to lie in a ventral position within the cerebral hemispheres. A large bundle of fibers, the *fornix,* containing the connections to the septal and hypothalamic regions, traces the path of this migration. These relations are indicated diagrammatically in Fig. 50.

In its position within the cerebral hemispheres, the hippocampus lies close to the olfactory cortex, as is shown in Fig. 50. Traditionally, therefore, the hippocampus was considered an olfactory structure. This concept has been modified by the discoveries that there are no direct olfactory bulb fibers to the hippocampus (Brodal, 1947) and that the hippocampus is well developed in such animals as the dolphin, which lack olfactory bulbs altogether. These discoveries had the useful effect of removing the term *rhinencephalon* from the literature as a catch-all term for those various and sundry nether parts of the brain that were

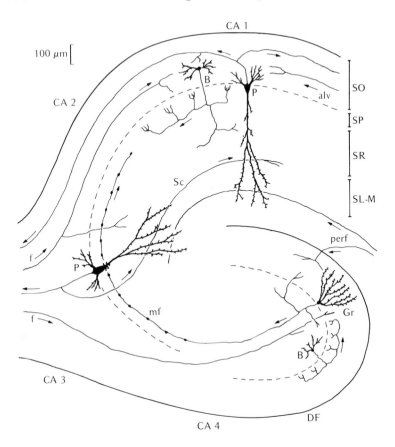

FIG. 56. Neuronal elements of the hippocampus.

Inputs: right, fibers in the perforant (perf) and alvear (alv) pathways; *left,* fibers arriving through the fornix (f); *below,* mossy fibers (mf) from dentate fascia.

Principal neuron: pyramidal neuron (P); recurrent Schaffer collateral (Sc).

Intrinsic neuron: basket cell (B).

The hippocampal regions are indicated by CA 1-4.

Neuronal elements of the dentate fascia are shown at DF.

Inputs: from the right, perforant (perf) pathway; from the left, fibers arriving through the fornix (f).

Principal neuron: granule cell (Gr) gives rise to mossy fibers (mf).

Intrinsic neuron: basket cell (B).

not obviously related to other systems. But any field abhors a terminological vacuum, and the term *limbic system* has come to be used generally for these erstwhile rhinencephalic domains. These developments have had the effect of denying any olfactory function whatsover to the hippocampus, which seems, in the light of recent studies, to be too extreme a position. Recent work indicates that the hippocampus receives input from several sensory modalities: visual, auditory, somatic, and, when present, olfactory.

Another reason for the demotion of olfaction has been the long queue of other candidates. The anatomical position of the hippocampus makes it a key structure, a head-ganglion, of the limbic system. The primordial connections to the hypothalamus indicate a close involvement in such functions as endocrine control and the expression of emotional states. There has also been much interest in the role of the hippocampus in memory and learning. Finally the proclivity of the hippocampus to generate rhythmic activity may be mentioned, as well as its implication, thereby, in certain seizure states of clinical interest.

Apart from its relevance to these aspects of behavior, the hippocampus has been of interest to neuroanatomists and electrophysiologists because of its remarkably stereotyped internal structure. For this reason alone, it warrants an important place in the study of principles of synaptic organization in the brain. As we shall see, the results of recent studies provide useful comparisons with other regions of the brain. In addition, they provide the initial steps toward an understanding of the neural basis of the behavioral functions of the hippocampus.

NEURONAL ELEMENTS

The hippocampus forms a long cylinder, opened at one side; in cross section it has a C-shape as shown in Fig. 56. The dentate fascia has a similar form, its open face abutting onto the lower lip of the hippocampal cortex. Within the hippocampus, there are gradations of internal structure along the longitudinal axis. Also, as one passes along the circumference of the C, the structure varies in important respects. In one terminology (Lorente de Nó, 1934), there is a sequence of CA 1, 2, 3, and 4, as shown in Fig. 56. In another terminology (Cajal, 1911), there is a regio superior (roughly CA 1) and regio inferior (CA 2 and 3).

It can be gathered from these remarks that the internal structure of the hippocampus, although stereotyped, is by no means simple. The complicated geometry and the gradations of internal structure require that any description be based on an oversimplification, particularly if variations of species are further borne in mind. For the purpose of accurately describing the neuronal elements, we will at least retain as much of the geometry as possible. This is done in Fig. 56; the diagram is adapted from Cajal (1911) and Lorente de Nó (1934) and is drawn to the same scale as the diagrams of other brain regions.

It should be noted that because of the migratory contortions of the hippocampus, the deep layers abut on the surface of the lateral ventricle, whereas the superficial layers are covered by the infolding of the dentate fascia. The student must not be deceived by these appearances; the former are analogous to the deep layers, and the latter to the superficial layers, of other cortical regions.

INPUTS   One of the main inputs to the hippocampus comes from the entorhinal cortex. As shown in Fig. 50, this is a nearby cortical region of the temporal lobe. It is itself a complicated region, receiving as it does inputs from several sensory modalities (van Housen, Pandya and Butters, 1972) and from certain regions of the brain, the cingulum in particular. The fibers mediating these inputs may be regarded as *afferents* in the sensory sense of the term.

The fibers from the entorhinal cortex enter the hippocampus by two pathways. One of these is the so-called *perforant* pathway. These fibers pass through ("perforate") intervening regions (the subiculum) en route; as shown in Fig. 56, they enter the most superficial layer of the hippocampus, pass through region CA 1, and terminate in region CA 3 (Hjorth-Simonsen and

Jeune, 1972). The other part of the entorhinal input arrives through the so-called *alvear* pathway; these pass into the ventricular (alvear) layer and terminate in region CA 1.

Other inputs to the hippocampus come through fibers from the septal region and through commissural fibers from the hippocampus of the opposite side. These arrive through the fornix and distribute through the middle layers of the hippocampus. They are perhaps analogous to the *central* inputs of other regions we have considered.

Finally, the input to the hippocampus from the dentate fascia may be noted. This arrives through the so-called *mossy fibers*, which are distributed in a thin sheet within the hippocampus (see Fig. 56). They will be further described in our discussion of the dentate fascia.

PRINCIPAL NEURON The output from the hippocampus is carried in axons that arise from one type of cell, the *pyramidal neuron*. The cell bodies of the pyramidal neurons are arranged in a thin sheet about 300 μm below the ventricular surface. As the name implies, the cell bodies have a pyramidal shape, imparted to them by their apical and basal dendritic trunks. The cell bodies vary in size, being 20-40 μm across at the base and 40-60 μm in height. Each has a stout apical dendrite, 5-10 μm in diameter, at its origin. As shown in Fig. 56, these dendrites are directed in radial array toward the "surface" (see preceding discussion) of the cortex; they reach lengths of 500-1000 μm. The apical dendrite gives off several side branches and typically divides into two or more terminal branches within the most superficial layer (stratum lacunosum-moleculare).

It is notable that there is a gradation of cell body size and apical dendrite configuration as one goes from region CA 1 to CA 3: as can be seen in Fig. 51, the cell bodies become larger, and the apical dendrites become rather shorter, stouter, and more irregular and profuse in their branching patterns. The largest neurons of CA 3 are, in fact, referred to as *giant pyramidal cells*. Thus, although the hippocampus has one type of output neuron, there is a distinct regional variation within that type; this is in contrast to the cerebellum and olfactory bulb but similar to the regional variation in the retina and the cerebral cortex.

The basal dendrites arise as several trunks, 3-6 μm in diameter. They are oriented toward the ventricular surface and arborize extensively over a field some 200-300 μm in diameter. A striking feature of both basal and apical dendrites is their liberal investiture with *spines*.

The axons of the pyramidal neurons are directed toward the ventricular surface, where they form the *alveus*, whence they pass laterally to form a sheet, the *fimbria* (see Fig. 56). Within this ambit, they give off several types of collateral. Some collaterals are short and local in character, simply recurring to the immediate vicinity. A special type is the so-called *Schaffer collateral*, which arises principally from the pyramidal neurons in regio inferior. They arise a short distance (several hundred microns) from the origin of the axon as a thick myelinated branch. According to the Golgi studies (Schaffer, 1892; Cajal, 1911; Lorente de Nó, 1934), these collaterals pass across the hippocampus within the same *axial* segment into the apical dendritic layer of regio superior, as is shown in Fig. 51. These collaterals terminate within the middle layer (stratum radiatum) of the regio superior. Another specific type of collateral from the axons of regio inferior neurons is oriented along the *longitudinal* axis of the hippocampus to terminate in the middle layer of the regio inferior. These were termed *longitudinal association fibers* by Lorente de Nó (1934). The question of whether the Schaffer collaterals and the longitudinal association fibers form two distinct types, or are at two extremes of a continuum, has been discussed by Hjorth-Simonsen (1973).

From the fimbria, the pyramidal axons gather to form the *fornix*, through which they pass in a long arc anteriorally and then ventrally to their projection sites deep in the brain. The main sites are the mammillary bodies of the *hypothalamus*, the anterior nuclear group of the *thalamus*, and the *septal region*. Together with the hippocampus, they form part of the limbic system, which, as already mentioned, is implicated in a variety of behavioral mechanisms. In addition to these projections through the fornix, there is evidence from recent studies (Hjorth-Simonsen, 1973; Andersen, Bland, and Dudar, 1973) for a projection of the pyramidal axons of CA 1 to the nearby subicular cortex, as shown in Fig. 56.

The significance of these projections would take us far afield and involve us in much putative data and speculative schemes. Let us simply note, as an example, the fact that, through the projections to the hypothalamus and thalamus, a relay is made to the cingulate cortex, which in turn projects back to the hippocampus, albeit indirectly, through the subiculum. Thus, it can be seen that the hippocampus is involved in long feedback circuits. If we recognize the fact that different parts of the hippocampus project to different parts of the regions in these circuits, we begin to realize that the complexity of the relations between the hippocampus and the rest of the brain exceeds our comprehension of their significance. For an excellent guide to the connections of the hippocampus, and their relation to the limbic system, the reader is referred to the account of Nauta (1971).

INTRINSIC NEURONS   There is a variety of neurons whose axons remain within the hippocampus. By this token, they may be regarded as *short-axon cells;* on the basis of the irregular shapes of their cell bodies and dendrites, they are referred to as *polymorphic cells*. An example of this type is shown in Fig. 56(B); its cell body is located in stratum oriens. The cell bodies range in diameter from 15-30 μm; each gives rise to several relatively stout dendrites, 3-6 μm in diameter. The dendrites are notable in that they branch sparingly and have very irregular orientations; the dendritic surface, although knobbly, has few spines. The dendrites attain lengths of several hundred microns. Thus, as intrinsic neurons go, these are rather large.

As Fig. 56 shows, the axon of this cell type follows a complicated course. It first ascends through the pyramidal cell body layer to the middle layer (stratum radiatum), where it undergoes several divisions. Some of the branches terminate within the middle and superficial layers, whereas others recur to the layer of pyramidal cell bodies. Here they ramify extensively, forming clusters of terminals around individual pyramidal cell bodies and dendrites. There is a resemblance in outward form to the terminations of basket cells in the cerebellum (Chapter 8); this resemblance extends also to the basket terminations in the neocortex (see Chapter 12).

Other types of intrinsic neuron are found at other levels in the hippocampal cortex. The shapes of the cell bodies and dendritic fields of these intrinsic neurons appear to reflect the orientations of surrounding structures. For example, near the ventricular surface (alveus) and at the other surface (stratum lacunosum-moleculare), there is a horizontal orientation, whereas in the middle layer (stratum radiatum), there is a vertical orientation. In general, the axon of a superficially located cell distributes deeper in the cortex, and vice versa. It is interesting to note that Cajal (1911) mentioned an intrinsic neuron that lacks an axon, in the stratum lacunosum-moleculare. These and other varieties of intrinsic neuron are not included in Fig. 56 for the sake of simplicity.

CELL POPULATIONS   Counts of the number of fibers in the fornix on one side have yielded estimates of 500,000 in monkey (Daitz and Powell, 1954), 1,200,000 in man (Powell, Guillery, and Cowan, 1957; see also Brodal, 1970). This is only a rough indication of the number of principal neurons (i.e., pyramidal cells) in the hippocampus because the axons branch in their course within the fornix to give rise to commissural fibers. It may be noted that this is a relatively large population of principal neurons, larger than that of the olfactory bulb (50,000) or the entire ventral horn (200,000), about the same as that of the retina and the pyramidal tract from the cerebral cortex (see Chapter 12), but smaller than that of the cerebellum (7,000,000). No figures are available for the numbers of input fibers to the hippocampus, so no estimate can be made of the input-output ratio; one can guess that, in view of the large number of output neurons, the ratio might be relatively low. Similarly, no estimates are available for the numbers of intrinsic neurons. It is stated (Cajal, 1911) that they are much fewer in number than the pyramidal cells. This is offset by the very large divergence from one basket cell onto the pyramidal cells. It has been estimated that one basket cell in the hippocampus establishes contact with 200-500 pyramidal cells (see Eccles, 1969). This is more than an order of magnitude greater than the comparable estimate for the basket cells in the cerebellum. The over-all I:P ratio might be estimated very roughly to be of the order of 1-10:1.

DENTATE FASCIA   The main *input* to the dentate fascia is through the fibers of the *perforant pathway* (see Fig. 56); most of these fibers, in fact, terminate in the dentate. It is especially notable that the entorhinal cortex projects to both the hippocampus and the dentate fascia over this pathway. Other inputs come from collaterals of pyramidal neurons in regio inferior of the hippocampus on the same and opposite sides and from septal nuclei through fibers in the fornix.

The *output* from the dentate fascia is through the axons of so-called *granule cells*. As shown in Fig. 56, their cell bodies are arranged in a thin sheet about 200 μm below the surface. The cell bodies are only about 10 μm in diameter. From the superficial aspect of the cell body arise several dendritic trunks (each several microns in diameter); these trunks course through the superficial (molecular) layer and branch and terminate near the surface. The dendrites are richly invested with spines. In outward form these cells are not unlike smaller versions of cerebellar Purkinje cells. This gives us the opportunity to once again point out that the term *granule* has no necessary significance, either anatomical or physiological, in comparing neurons of different regions. These granule cells are not of the same morphological type as their namesakes in either the olfactory bulb or the cerebellum.

The axons of the granule cells give off several collaterals within the dentate fascia. They then gather to form a band (or, more correctly, a sheet) of fibers that come to lie within the hippo-campus just superficial to the layer of pyramidal cell bodies (see Fig. 56). Along their lengths these axons have large varicosities, and they end in bulbous swellings; this has earned them the name of *mossy fibers*. Again, there is no relationship to the mossy fibers of the cerebellum.

The *intrinsic* neurons of the dentate fascia include various kinds of *short-axon cells*. They are found at all levels and have shapes and sizes not unlike many of the polymorphic cells of the hippocampus. The main type has a cell body situated in the deep layer; as can be seen in Fig. 56, the axon ascends to the superficial layer and then recurs to the layer of granule cell bodies, where it ramifies and terminates in clusters of terminals. These cells have been regarded as analogous to the basket cells of the hippocam-pus (cf. Eccles, 1969), although this is a comparison Cajal (1911) did not himself make. It is interesting to note that, here and there, are found neurons that have a pyramidal form, with apical and basal dendrites, but whose axons ramify locally (Cajal, 1911). The fact that such a neuron is, by definition, a short-axon cell, is a reminder that the term *pyramidal* is descriptive of outward form only.

No estimates are available for the numbers of input fibers or of principal neurons in the dentate fascia. It seems safe to guess that the number of principal neurons is rather high. Similarly, there are no estimates of the population of intrinsic neurons, but they are certainly less numerous than the principal neurons.

SYNAPTIC CONNECTIONS

By virtue of the lamination of fibers in the hippocampus and den-tate fascia, it has been possible to obtain reliable evidence in both these regions for the identification of connections made by differ-ent types of synaptic terminal (see Hamlyn, 1963; Blackstad, 1967; Nafstad, 1967; Gottlieb and Cowan, 1971). The main types are illustrated in Fig. 57.

The input fibers in the perforant pathway make axodendritic

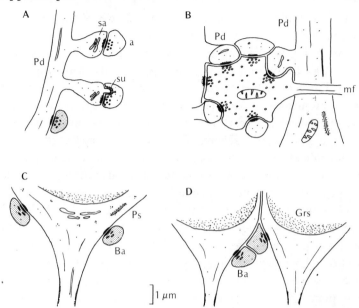

FIG. 57. Synaptic connections in the hippocampus. (A) Axodendritic connections onto a pyramidal cell dendrite (Pd); note the spine apparatus (sa) and spinule (su). (B) Axodendritic connections from a mossy fiber terminal (mf) onto the proximal dendritic shafts and spines of pyramidal cells (Pd). (C) Axosomatic connections onto the cell body of a pyramidal neuron (Ps) from basket cell axons (Ba). [(A)-(C) after Hamlyn, 1963; Westrum and Blackstad, 1962; Blackstad, 1967; Gottlieb and Cowan, 1971.)

Synaptic connections in the dentate fascia. (D) Axosomatic connections from basket cell axons (Ba) onto cell bodies of granule cells (Grs). (After Laatsch and Cowan, 1966.)

synapses onto the spines of the apical dendrites of the pyramidal cells. As shown in Fig. 57(A), these are mainly type I chemical synapses, with asymmetrical membrane thickenings and round vesicles. The spines are rather small, with diameters of about 1 μm or less; the larger ones contain a spine apparatus. Many of the spines have a "spinule" protruding into the presynaptic terminal, as shown in Fig. 57(A).

Next to be described are the input connections made by the

dentate granule cells onto the pyramidal cells. As already mentioned, these are by way of the mossy fibers and terminals. As shown in Fig. 57(B), the terminals are some 3-6 μm in diameter, in contrast to the thin axons (less than 0.5 μm) from which they arise. The terminals have tortuous shapes, with many protrusions and invaginations, which accommodate the shafts and spines of the apical dendrites near their origin from the pyramidal cell bodies. It is particularly striking that a terminal may entirely surround a dendrite, as is shown in Fig. 57(B). Each mossy terminal has synaptic connections onto numerous dendrites and spines. The contacts are type I chemical synapses. In addition to these large terminals, the mossy fiber has numerous swellings along its course; these swellings are the sites of *en passant* type I synapses onto the pyramidal cells. The swellings, like the terminals, are packed with vesicles ranging in size up to 2000 Å (0.2 μm) diameter, some of which contain dense cores [see Fig. 57(B)].

A point of some interest is that the mossy fiber terminal has been shown to have an exceptionally high content of zinc (von Euler, 1962; McLardy, 1963). Special stains have demonstrated the presence of this metal throughout the extent of the mossy fibers and within the granule cells of the dentate fascia from which they arise. This intriguing finding is little understood, beyond the speculation that the zinc is associated with some protein or enzyme.

A third type of connection within the hippocampus has been identified within the layer of pyramidal cell bodies. Here are found terminals with contacts onto the cell bodies; these axosomatic synapses are type II chemical contacts [see Fig. 57(C)]. It is believed that these synapses are made by the axon terminals of the basket cells. The majority of synapses onto the somata of pyramidal cells have been reported to be of this type (Gottlieb and Cowan, 1971). These contacts are not numerous; Gottlieb and Cowan have estimated that they occupy less than 5% of the surface area of the pyramidal cell bodies. It is important to note that these terminals, and their synaptic contacts, stand in contrast to the highly specialized terminal complex made by a basket

cell axon around the cell body and initial axon segment of a Purkinje cell in the cerebellum (cf. Chapter 8). The evidence at the
synaptic level, therefore, suggests that great caution be used in
implying a similarity in structure and function between the basket terminals in the cerebellum and the hippocampus.

We have thus far described three of the main types of synaptic
connection within the hippocampus. There are many more.
Among them should be mentioned the connections made by the
Schaffer collaterals from the pyramidal cell axons; these are made
within the middle layer (stratum radiatum) and have been reported to be mostly type I axodendritic synapses made onto the
branches and spines of the apical dendrites. Elsewhere in this
layer, most of the synapses are type I onto the dendritic spines;
these synapses come from the afferent fibers from the septum
and from commissural fibers from the opposite hippocampus. A
small number of type II chemical synapses have been found onto
the apical dendrites in the "superficial" zones. In "deeper" layers,
type I synapses are present on the basal dendrites of the pyramidal cells; type I synapses are also presumed to arise from fibers
from the septum and commissure. For the sake of simplicity we
will not depict these various connections here.

In the *dentate fascia*, we have already noted that the perforant
fibers make type I synapses onto the spines of the granule cells
(Laatsch and Cowan, 1966). These connections resemble rather
closely their counterparts in the hippocampus—the spines are
small (0.5-1 μm in diameter); many spines have a spinule protruding into the presynaptic terminal, similar to what is seen in the
hippocampus. Most of these connections from the perforant
pathway are made onto the spines of intermediate and distal
branches of the granule cell dendrites. Proximally, connections
onto large dendritic branches and onto stubby spines come from
association and commissural fibers; these are also type I chemical
synapses.

On the granule cell bodies are found terminals that make
type II synapses, as is shown in Fig. 57(D). Electron-microscope
studies have, in addition, revealed several unusual features of the

granule cell bodies. These include the fact that there is only a thin rim of cytoplasm around the nucleus, that there are no distinct Nissl bodies, and that the cell bodies are packed so tightly that much of the surface membrane of a cell is directly apposed to that of its neighbor. These features are indicated only very sketchily in Fig. 57(D). In addition to these findings regarding the cell bodies, it has also been found that the mossy axons arising from them form bundles or sheets within which their surface membranes are also directly apposed, a situation not unlike that of the unmyelinated olfactory nerve fibers (Chapter 6) and the parallel fibers of the cerebellum (Chapter 8). The obvious possibility for "ephatic" interactions through the appositions of the cell bodies and the axons of the granule cells has been noted (Laatsch and Cowan, 1966) (cf. Chapter 2).

BASIC CIRCUIT

The organization of the hippocampus, together with that of the fascia dentata, is summarized in the basic circuit diagram of Fig. 58. It is well perhaps to begin by acknowledging the problem of simplifying the geometry of the hippocampus to fit with the conventions used in the diagrams for other parts of the brain. The diagram follows the usual practice in the literature, of reducing the hippocampus to its two main regions, regio superior and inferior, and placing by its side a single representation of the dentate region (see, for example, Hamlyn, 1963; Gottlieb and Cowan, 1973). As in most diagrams in this book, the main afferent input arrives from above and the output of the region is below. It is to be hoped that we have not, like Procrustes, mutilated the traveler by fitting him to this bed.

By way of brief summary, the perforant pathway connects to the distal dendrites of the hippocampal pyramidal cells in regio inferior. The alvear pathway connects to the basal dendrites of cells in regio superior. The third input pathway, through the fornix, distributes as indicated by the dotted line.

Within the hippocampus, two main circuits are organized in

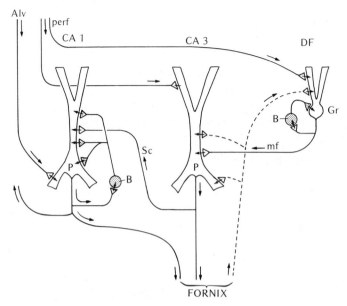

FIG. 58. Basic circuit diagram for the hippocampus and dentate fascia. Abbreviations as in Fig. 57.

relation to the recurrent collaterals of the pyramidal neurons in the two main regions. One is formed by the Schaffer collaterals from the pyramidal neurons of regio inferior to regio superior. The other is formed by the collaterals of regio superior neurons to basket cells, which feed back to the cell bodies and the dendrites of the pyramidal neurons in that same region, as is shown in Fig. 58.

The dentate fascia receives its main input from the perforant pathway, as shown in Fig. 58; other inputs come from the septal and commissural fibers in the fornix. Within the dentate is postulated a feedback circuit from granule cell axon collaterals to basket cells and back onto the granule cell bodies. The output from the dentate fascia, through the mossy fibers of the granule cells, is directed to the pyramidal neurons of regio inferior of the hippocampus.

In assessing these organizational patterns, the first fact to be noted is that all the connections are axosomatic or axodendritic; there are no dendrodendritic or otherwise oriented synapses. Furthermore, there are no synaptic triads, as we have defined them, in these circuits; the perforant fibers, for example, make connection only to the principal neurons and not to any intrinsic neurons. Whether this is true of inputs arriving through the fornix is not known.

With regard to intrinsic pathways, there is evidence, to be discussed in the next section, that the intrinsic neurons, like their counterparts in other parts of the brain, are inhibitory. All the other connections in the hippocampus and dentate fascia have been reported to be excitatory, as will be described in the next section. Of particular interest in this respect are the Schaffer collaterals, which, as Fig. 58 illustrates, provide for excitatory feedback from pyramidal neurons onto other pyramidal neurons.

Most of these characteristics of hippocampal and dentate organization have counterparts in the olfactory cortex (Chapter 10). They stand in contrast to olfactory bulb and retina, in which dendrodendritic synapses are prominent, synaptic triads are the basis of synaptic organization, and principal neurons feed back onto intrinsic neurons but not onto each other. The reader may further study the points of similarity and difference with other regions—ventral horn, cerebellum, and thalamus; we will consider some of these points later. Comparison with the neocortex will be made in Chapter 12.

The diagram in Fig. 58 emphasizes that a horizontal and vertical framework provides the main scaffolding for the connections in the hippocampus. In this respect, the framework is similar to that of other cortical and laminated regions of the brain. The recognition of this framework is important, since certain implications immediately follow for the functional organization of the hippocampus.

In the *horizontal* plane, there are clearly different regions as one moves around the circumference of the hippocampus, and this means that these regions form narrow strips or segments in the longitudinal axis of the hippocampus. These regions grade into each other much more gradually than the diagram of Fig. 58 indicates; there are gradations in the longitudinal axis as well. It is important to note that there are *multiple bases for regional parcellation*; there are specific inputs to different regions, specific intrinsic circuits within and between different regions, and specific projection sites from different regions; these are all illustrated in Fig. 58.

In this respect, the hippocampus is more complicated than most other regions of the brain. In the olfactory bulb, for example, there is no evidence for parcellation on any of these bases (Chapter 6). In the cerebellum, different parts have different input and output connections, but the intrinsic circuits, basically, appear similar, and there are no obvious sequences from one part of the cerebellum to another (Chapter 8). The neocortex, of course, is characterized by multiple parcellation on all these bases; we need only mention, as an example, the differences between the motor cortex and the visual cortex (Chapter 12). By way of speculation, one may ask whether it is possible that the regions of the hippocampus are analogous to motor and sensory areas of the neocortex? Or must we conceive of entirely different central or behavioral "modalities"? In reply to such questions, the hippocampus only grins back like a Cheshire cat.

In the horizontal plane there is at any rate a clear implication of a *sequencing of activity* in several of the circuits. Prominent among these are (1) the mossy fiber input from dentate fascia to regio inferior and (2) the Schaffer collaterals from regio inferior to regio superior. This sequencing of activity in the horizontal plane is similar to, though more rigid than, that in the olfactory cortex (Chapter 10). It is also similar to that in the cerebellum (Chapter 8), with the qualification that, in the cerebellum, the sequencing through the parallel fibers is bidirectional, not unidirectional as in olfactory and hippocampal cortex. Some functional implications of the organization of these circuits in the hippocampus will be discussed later in relation to synaptic action and dendritic properties.

The circuits that have been described above are notable in being oriented transversely across the longitudinal axis of the hippocampus. In recent experiments, these circuits have been mapped by anatomical (Blackstad, 1967; Hjorth-Simonsen and Jeune, 1972), histochemical (Storm-Mathisen and Blackstad, 1964), and physiological means (Anderson, Bliss, and Skrede, 1971). From these studies, it has been concluded that the hippocampus is organized into parallel transverse lamellae. The diagram of Fig. 58 may, therefore, be taken to represent the organization of one lamella.

In the *vertical* axis, there is a layering, or lamination, of inputs to the output neuron (pyramidal neuron). The attempt has been made in Fig. 58 to indicate the vertical locus of various inputs onto the dendritic trees of the hippocampal (and dentate) neurons. This lamination has traditionally been regarded as the outstanding feature of hippocampal organization (cf. Lorente de Nó, 1934), more distinct than in any other region of the brain. We have already noted, however, the lamination of inputs to the dendrites of granule cells in the olfactory bulb (Chapter 6) and the olfactory cortex (Chapter 10). It should be clear that, in one form or other, this is a common feature in the organization of most regions of the brain. We will discuss its significance for neocortical organization in Chapter 12. The interpretation of the functional significance of the inputs at different vertical levels in the hippocampus is dependent, as elsewhere, on a knowledge of the dendritic properties of the pyramidal cells, which will be discussed later in this chapter.

Let us finally discuss briefly the dentate fascia. As the diagram of Fig. 58 indicates, this region consists essentially of an input-output relay with an intrinsic feedback circuit. The diagram also indicates the special position of this region with regard to its input and output. The essence of this position was recognized long ago by Cajal (1911), and we can do no better than to quote from him:

> The granules . . . are not long-axon neurons . . . but rather special corpuscles, of which the semi-long axon is charged . . .

with carrying to the bodies and the protoplasmic shafts [i.e., den-
drites] of the large pyramidal cells the olfactory excitations that
they receive from the temporoammonic pathway.

We see, therefore, that the granule cells provide a special relay
along the entorhinal input pathway to the large hippocampal pyramidal cells. This, together with the fact that the granule cells
have only this one output, suggests that the dentate may be
analogous to the thalamus and to the granule cells of the cerebellum, in that all three of these regions perform some special
function of staging or preprocessing of an input to a specific
cortical region.

SYNAPTIC ACTIONS

Even more so than the cerebellum, the hippocampus is at a re-
move from the major sensory input pathways in the brain, as
well as from the final motor outflow through the motoneurons.
The neurophysiologist is, therefore, faced with a region in which
there are few obvious or identifiable characteristics of the inputs
or outputs that would provide starting points for the analysis of
synaptic actions. Indeed, in such a situation, the problem gets
stood on its head, and the analysis of synaptic actions becomes
part of the evidence for what those characteristics might be. At
any rate, the input and output pathways are clearly separated, so
that electrophysiological analysis of synaptic actions can be carried out as effectively as in the spinal cord and the olfactory bulb
(see Kandel, Spencer, and Brinley, 1961; Andersen, Blackstad,
and Lømo, 1966a; Andersen, Holmquist, and Voorhoeve, 1966b).

The responses of the hippocampal pyramidal neurons to single
volleys in the different input pathways are illustrated in Fig. 59.
In (A) there is an intracellular recording of the response to a
volley originating in the olfactory bulb. It consists of a long-
latency, slow-depolarizing wave, which has been ascribed to an
EPSP elicited over a polysynaptic pathway from the bulb. Presumably this route passes from the olfactory bulb to the ento-

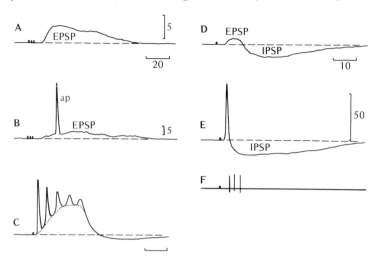

FIG. 59. Synaptic actions in the hippocampus. (A) Intracellular response of a pyramidal cell to a volley from the olfactory bulb. (After MacLean et al., 1970.) (B) Response to a volley from the septum, through fibers in the fornix (MacLean et al). (C) Response to a strong volley in the perforant pathway. (After Kandel and Spencer, 1961.) (D) Response to a volley in the fornix. (After Kandel et al. 1961.) (E) Antidromic response to a volley in the de-afferented fornix (Kandel et al.). (F) Extracellular unit recording of a presumed basket cell response to a volley in the fornix. (After Andersen et al., 1964.)

rhinal cortex (Fig. 50), thence, through one or more synaptic relays, to the hippocampus via the perforant pathway. A notable feature of the response is that it does not lead to impulse firing; the synaptic potential remains subthreshold for impulse initiation in the pyramidal neuron. This has been interpreted as indicating that the EPSP is set up in the distal branches of the apical dendrites where the fibers of the perforant pathway terminate (Yokota, Reeves and MacLean, 1970). These findings indicate that there is in fact an olfactory input to the hippocampus, but that its synaptic action appears to be limited to subthreshold effects. The suggestion has been made that these synaptic actions are comparable to conditional stimuli that might have a role in memory and learning (MacLean, 1972).

A single volley from the septal region elicits in a pyramidal neuron the response shown in Fig. 59(B). The long-lasting, low-amplitude depolarizing wave has been ascribed to an EPSP and is associated with the discharge of one or two impulses (Yokota et al., 1970; see also Andersen et al., 1966a,b). MacLean (1972) has pointed out that, through the septum, the hippocampus receives input from the hypothalamus, which is involved in aversive, appetitive, visceral, and humoral reactions of an unconditional nature. In his view, the septal inputs are comparable to unconditional stimuli, insofar as they are themselves capable of causing pyramidal impulse discharge. Whatever the interpretation, the fact that synaptic excitation is so effective over this route, in contrast to the inability of large amplitude EPSP's to elicit spikes over the route from the olfactory bulb (described above), appears to be an important aspect of synaptic integration in the hippocampus.

A distinctive type of discharge in the pyramidal neuron is seen in response to a strong volley in the perforant or the septal (fornix) pathways. This consists of a rapid discharge of several spikes that arise from an intense depolarizing wave [Fig. 59(C)]. As the wave builds up, the spikes decrease in amplitude, until the discharge is terminated. This has been termed an *inactivation burst* (Kandel and Spencer, 1961), in analogy with the similar discharges (complex spikes) of Purkinje cells in the cerebellum (see Chapter 8). In the hippocampal neuron, it seems that the spikes are generated in response to the prolonged membrane depolarization and that the termination of the spike discharge is directly determined by the amount of excessive depolarization, which gradually "inactivates" the spike-generating mechanism. Such a mechanism depends on the membrane properties of the pyramidal neuron rather than the synaptic circuits in which the neuron is involved; see the discussion of a similar question with regard to the complex spikes in the Purkinje cells of the cerebellum and the neurons of the inferior olivary nucleus (Chapter 8). This tendency to discharge in bursts has been taken to reflect the highly excitable nature of the pyramidal neurons, a property that has important implications for the generation of rhythmic activity in the hippocampus (see below).

INHIBITORY PATHWAYS The analyses of synaptic actions and synaptic morphology are consistent, as far as is known, with the view that all the external inputs to the hippocampus are excitatory; in addition, there is an internal pathway for re-excitation through the Schaffer collaterals. Against these numerous excitatory inputs is a rather small population of intrinsic neurons for inhibitory actions. Despite this, electrophysiological experiments give ample evidence of strong inhibitory actions within the hippocampus, in response to activation over any of the input pathways. A typical recording is shown in Fig. 59(D), in which the response of a pyramidal neuron to a volley in the fornix is shown. The response consists of an initial depolarization followed by a long-lasting hyperpolarization, which have been interpreted as an EPSP and IPSP, respectively (Kandel et al., 1961). This is another example of the excitatory-inhibitory sequence that is characteristic of synaptic actions engendered by an input volley in the principal neurons of many regions of the brain (see Purpura, 1967). The inhibitory period is often followed by a rebound excitation, similar to that in neurons in the thalamus (Chapter 9).

That the EPSP is due to connections of the input fibers [septal afferents in the case of Fig. 59(D)] has been shown by chronic transection of the fornix, which causes these fibers to degenerate. This leaves only the pyramidal axons in the fornix, and the response to a volley in these axons is simply an antidromic spike followed by the IPSP, as shown in Fig. 59(E).

The pathway for the IPSP has been deduced from experiments analogous to those carried out in the spinal cord. Microelectrode recordings from small units in the superficial layer of the hippocampus have revealed responses characterized by a repetitive discharge that begins within a millisecond or so of the antidromic spike, and that continues into the time period of the IPSP. An example of an extracellular recording from such a unit is illustrated in Fig. 59(F). In analogy with the Renshaw circuit in the spinal cord, it has been concluded that these units are basket cells and that there is a pathway from pyramidal axon collaterals to basket cells back onto the pyramidal cells (Andersen, Eccles and Løyning, 1964; Eccles, 1969). The sites of inhibitory input by the basket cells to the cell body and the apical dendrites of the pyramidal neurons have been indicated in the basic circuit diagram of Fig. 58.

Several points about this inhibitory circuit stand out. First, the synaptic connections, like those in the Renshaw circuit, are entirely axo-somatic/dendritic. This means that the inhibitory actions are diffuse, rather than local, as in the case of a dendrodendritic recurrent pathway (see Chapter 6); this diffuseness is accentuated by the small population of basket cells and the high divergence of axonal output from a single basket cell onto 200 to 500 pyramidal cells. Second, the position of this circuit indicates that it functions largely as a feedback mechanism, because it is not accessible to the feedforward input from the perforant and mossy fiber pathways. Third, and finally, the inhibitory potentials are of large amplitude (10-20 mV) and long time course (100-300 msec); this is in contrast to the synaptic potentials in motoneurons (Chapter 5).

In sum, therefore, the intrinsic (basket) neurons appear to provide for inhibitory control, at the level of pyramidal neuron output, that is powerful, long-lasting, diffuse, and largely of a feedback nature. It may well be that these are the properties necessary to oppose the powerful excitatory pressure of the hippocampal inputs. Note that the basket cell axons ramify in the transverse plane (Fig. 56), so that they also reflect a lamellar organization. The inhibitory field established within a lamella is reminiscent of the compartment of inhibition established by the widely ramifying axon of a Golgi cell in the cerebellum (Chapter 8).

RHYTHMIC ACTIVITY The hippocampus is notable for the prominent rhythmic potentials it generates. It might be thought, and indeed it was thought by the early workers, that the hippocampus would provide a model system for studying the generation of rhythmic activity. The subject, however, has become quite complicated, for it encompasses not only the vexed problem of the interpretation of summed extracellular potentials but also the tendency of the hippocampus to develop paroxysmal discharges and, finally, the relation of these discharges to epileptic seizure states. Our interest must be restricted to certain aspects relevant to synaptic organization.

The resting activity of the hippocampus takes the form of relatively slow waves (3-8/sec), the *theta rhythm*. The sequence of excitation-inhibition-rebound excitation in hippocampal pyramidal cells (Fig. 59) appeared to be an obvious mechanism for the generation of the theta rhythm, in analogy with the inhibitory phasing of activity in the thalamus (Chapter 9). The experimental evidence thus far, however, is not entirely consistent with this mechanism, and the subject requires much more investigation (see Green, 1964; Spencer and Kandel, 1968).

If rhythmic activity is forced on the hippocampus by repetitive stimulation (e.g., of the fornix), some remarkable properties of the synaptic actions onto pyramidal neurons are revealed (Purpura, 1967; Spencer and Kandel, 1968). These properties take the form of sustained shifts of the membrane potential, potentiation of EPSP's to large amplitudes, depression of IPSP's, and a long-lasting period in the aftermath of stimulation during which evoked EPSP's are much larger than normal. These findings may be taken to reflect a remarkable lability in the potency of excitatory and inhibitory synapses in the hippocampus, also notable for its long-lasting time course. These are, of course, the kinds of properties that are suspected of being crucial for learning and memory, as has already been discussed (Chapters 3 and 8). However, the task of proving that these mechanisms are operative in the hippocampus has barely begun. We have previously noted changes in potency of certain synapses onto motoneurons (Chapter 5), but the effects in the hippocampus are much more prominent and complex. Very similar effects have been found in the neocortex, however, and are illustrated more fully in Chapter 12.

An increase in the level of excitability of hippocampal neurons, either occurring naturally or under artificial conditions of electrical stimulation in an irritative focus, produces an unstable situation, which rapidly gives way to uncontrolled activity. This can take various forms; a common form consists of a large depolarizing wave in a pyramidal neuron of 30-40 mV, which generates a burst of impulses not unlike the burst response already described (Fig. 59). This is followed by a prolonged plateau of depolarization and discharge of impulses, which constitute the period of the epileptic seizure. The initial wave is termed a *paroxysmal depolarizing shift* (PDS) (Matsumoto and Ajmone-Marsan, 1964) and is currently under intensive investigation for the insight it can give into the mechanism that triggers a seizure. In one view, the PDS and subsequent discharge are due primarily to altered excitability of neuronal membrane (the "epileptic neuron"); in another view, they are due to abnormal activity circulating through the synaptic circuits (the "epileptic aggregate"). An excellent summary of recent research in this problem has been provided by Ayala, Dichter, Gumnit, Matsumoto, and Spencer (1973). The important point for present concern is that the analysis of the mechanism of this neurological problem—epileptic seizure—is being carried out precisely at the level of synaptic organization. Very similar problems are currently under investigation in the neocortex (Chapter 12).

DENTATE FASCIA   In the dentate fascia, a volley in the perforant pathway (the main input) elicits in the granule cells an EPSP-IPSP sequence similar to that in the hippocampal pyramidal cells. Unit recordings deep to the granule cells have provided evidence that the basket cells there respond with a discharge similar to that of the basket cells in the hippocampus. It has therefore been concluded that there is a feedback circuit from granule cell axon collaterals through basket cells back into granule cells, as in the similar circuit in the hippocampus, and in analogy with the Renshaw circuit in the spinal cord (Andersen et al., 1966a). Note that the inhibitory interneuron (basket cell) is not directly accessible to the afferent input in the perforant pathway; the dentate fascia and the hippocampus are similar in this respect.

Recently it has been found that the granule cells respond to repetitive stimulation in much the same way as do the hippocampal pyramidal cells, that is, with a potentiation of EPSP's, depression of IPSP's, and a long-lasting aftermath of potentiation; according to Bliss and Gardner-Medwin (1973), the potentiation may be detectable over periods of hours or even days. This is long enough to be implicated as a possible mechanism for information storage, but "whether or not the intact animal makes use in real life of a property which has been revealed by synchronous, repetitive volleys . . . is", as Bliss and Lømo (1973) wisely caution, "another matter." There is, at any rate, a remarkable matching of the properties of hippocampal and dentate cells in this respect.

DENDRITIC PROPERTIES

The hippocampal pyramidal neuron is dominated by its apical and basal dendritic trees, but, thus far, there is no electrotonic model to serve as a basis for assessing their functional properties. The evidence indicates that the dendrites are only postsynaptic in position, in analogy with the dendrites of neurons in the ventral horn, cerebellum, and prepyriform cortex, and in contrast to the olfactory bulb, retina, and thalamic neurons. This is, of course, negative evidence, and we must allow for the possibility of future revision. For the present, however, we may confine our attention to local postsynaptic integration within the apical dendritic branches and transfer through the trunk to the axon hillock.

ELECTRICAL PARAMETERS   A model for the electrotonic properties of the apical dendritic tree should start with some evidence about electrical parameters. It has been reported that the total resistance (R_N) of a pyramidal neuron is in the range of 10 MΩ (Spencer and Kandel, 1961a). This is larger by a factor of five to ten than the value for motoneurons; this has been taken to reflect, to a large extent, the smaller size of the hippocampal neuron, but it probably also reflects a somewhat higher specific membrane resistance (R_m) of the neuronal membrane.

In response to an intracellular current pulse the membrane potential changes with an over-all constant (τ_N) of about 10 msec. This is several times longer than that for the motoneuron. An accurate estimate for the specific time constant of the dendritic membrane would require an electrotonic model. Nonetheless, is has been concluded (Spencer and Kandel, 1961a, 1968) that the longer time constant reflects a fundamental difference in the nature of the dendritic membranes, in that the longer time constant provides a greater opportunity for temporal summation of synaptic potentials than does a short time constant. We will see that neocortical pyramidal cells share this property.

These studies provide some of the data needed to assess the electrotonic properties of the pyramidal neuron, but a complete model is not yet available. We can, however, estimate that apical dendrites, with their diameter of 5-10 μm, might have characteristic lengths in the range of 700-1000 μm (cf. Fig. 13). This is about twice the actual length of the apical dendritic trunk; i.e., the electrotonic length of a trunk is of the order of 0.5, which means that the electrotonic potential spread must be relatively effective over this length.

FAST PREPOTENTIALS We next consider the properties of the terminal branches of the trunks. What can be said about the spread of synaptic potentials in these branches? We may confine ourselves to noting that if their diameters are several microns, their characteristic lengths might be of the order of 300-500 μm (see Fig. 12). This is in the range of actual lengths of these branches, as can be seen in Fig. 56, so that the electrotonic lengths of the branches would be in the range of 1.

Taking this estimate together with that for the trunk, one obtains an over-all electrotonic length for the apical dendritic tree of 1.5-2, not too different from that obtained for the dendritic trees of motoneurons and many of the other neurons we have considered. In the hippocampal pyramidal neuron, however, the synapses of the perforant fibers onto the terminal branches, at the farthest possible distance from the axon hillock, contrasts with

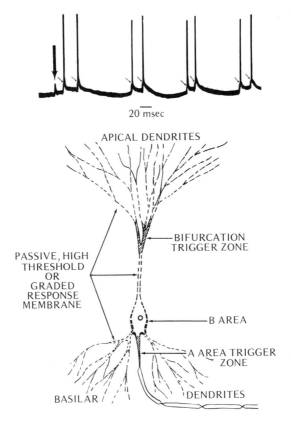

FIG. 60. Properties of hippocampal pyramidal cell dendrites. *Above,* intracellular recording from a pyramidal cell body of spontaneous action potentials. Large arrow, fast prepotential (FPP) in isolation; small arrows, inflections that indicate FPP's preceding large action potentials. *Below,* diagram to illustrate site of generation of FPP's in the bifurcation trigger zone of a dendritic tree. (From Spencer and Kandel, 1961b).

the case in the motoneuron, in which the inputs make synapses mostly over the middle range of the dendritic tree (see Fig. 23). This has posed the problem of how the postsynaptic response is transferred from these terminal branches to the axon hillock, in the face of what appears to be a considerable electrotonic decrement if the transfer is by passive means alone.

The answer to this problem seemed to be provided by the fact that in recordings from pyramidal neurons it is sometimes observed that the large spike arises from a small spike (see Fig. 60). These small spikes are seen during spontaneous firing and also in response to volleys in the input pathways, but they never precede an antidromic spike invading from the pyramidal cell axon. These small spikes are termed *fast prepotentials*. It is believed that they register a spike generated at an isolated patch of membrane somewhere in the apical dendritic tree.

The model developed by Spencer and Kandel (1961b) to explain the sequence of events is shown in Fig. 60. It envisages a patch of excitable membrane at the juncture of the peripheral branches with the dendritic trunk. Synaptic potentials spreading through the peripheral branches trigger the active membrane in the bifurcation zone to generate an all-or-nothing response analogous to, or identical with, an action potential. The action potential then spreads over the passive trunk membrane to the axon hillock to trigger the active membrane there. The sequence is somewhat similar to that in saltatory conduction (see Chapter 2) but without the intervening myelin. The dendritic trigger zone thus acts as a booster for transmitting peripheral dendritic input to axonal output. The model is similar to that for spike activity in the dendrites of chromatolytic motoneurons (Chapter 5) and cerebellar Purkinje cells (Chapter 8).

The capacity for active properties has several implications for synaptic organization. It seems that a distal input becomes much more effective vis-à-vis the axon hillock, indeed, possibly more effective than more proximal inputs whose responses spread only by passive means. It provides a more effective mechanism for overcoming inhibition along the proximal dendrites or at the cell body. It possibly increases the computational complexity of the dendritic tree, in line with Rall's comment (Chapter 8). Finally, it digitalizes the input, in contrast to the graded nature of synaptic responses spreading by passive means. The significance of these factors for the input-output functions of the hippocampus remains for further investigation.

12

NEOCORTEX

The term neocortex refers to the "new pallium", in the sense that it is the most recent of the cortical regions to differentiate in the phylogenetic series. It is a remarkable fact that throughout the vertebrate series most regions of the central nervous system are either clearly present (i.e., spinal cord, olfactory bulb, retina, cerebellum) or else have obvious or likely counterparts (i.e., thalamus and other brain-stem regions). This is also true of the palaeocortex (olfactory cortex) and, to a certain extent, of the archicortex (hippocampus). Some regions (i.e., retina) may be more elaborately developed in lower than in higher forms.

The neocortex, however, stands out as an exception. It is barely discernible in the brains of fishes and but ill-defined in amphibians and reptiles and even in birds. As a specific and well-differentiated area it is, in fact, virtually specific for mammals. Once established in the mammalian brain, it has evolved to such an extent as to account for the major part of the brain substance. It is not the only region to evince this capacity for overgrowth; the cerebellum achieves a similar overwhelming dominance in the brains of certain electric fishes, for example. Like the cerebellum, the neocortex has increased its volume not only by expanding outwards, but also by developing deep convolutions that greatly increase the surface-to-volume ratio; in fact two-thirds of the human cortex lies within the fissures of the convolutions. The olfactory cortex and hippocampus, in contrast, have retained their smooth surfaces.

The evolution of the neocortex has reached its greatest extent (we do not say climax!) in man. It is generally agreed that the capabilities that set man apart from other mammals, and mammals from other vertebrates, have their basis in the large population of cortical neurons, the complexity of interactions with other parts of the brain, and the complexity of interactions within the cortex itself. These factors lead directly to the study of synaptic organization as the necessary basis for understanding the particular contribution of the cortex to mammalian and human behavior.

Having said this, we must face the fact that few regions of the brain seem more designed to frustrate the analysis of synaptic organization than the neocortex. Two very obvious reasons can immediately be mentioned. One reason is that the fibers that carry the inputs and outputs arrive and depart through the depths of the cortex, where they are inextricably intermingled. Because they do not form separate and discrete bundles (as in most of the other regions we have studied) electrophysiological analysis is very difficult. The other reason is that, with the exception of those parts of the cortex that receive direct input from specific sensory pathways, the kinds of input information coming into the cortex are difficult to characterize. Much the same problem applies to the output side of the cortex, too. Consider the motor cortex. Although the parts of the cortex that send fibers directly (through the pyramidal tract) to the spinal motoneurons (as shown in Fig. 1) are obviously concerned with controlling movement, we have only a limited understanding, after a century and more of investigation, of the nature of that control in relation to the behavioral repertoire of mammals.

Consider, finally, the areas not frankly motor or sensory, the so-called association areas. It is obvious from the palaeontological record that it is these parts of the neocortex that have evolved the most in primate evolution and, therefore, must be most responsible for those capabilities that characterize human behavior— speech, thought, memory, etc. Yet it is precisely these areas that are the most difficult to analyze in terms of input, output, and intrinsic organization.

If knowledge of synaptic organization is necessary for understanding the neural basis of cortical function, study must begin
with the simplest part. As already indicated, the most readily
analyzed parts are those receiving inputs through the specific
sensory pathways that relay through the thalamus. Our focus will
therefore be on these regions. The aim is to sketch in only the
simplest outlines of organization that will have some generality
for the neocortex as an entity and will serve, at the least, as an
introduction to this most challenging of central synaptic systems.

NEURONAL ELEMENTS

By virtue of the outward expansion of the neocortex, the fibers
to it arrive through the depths, and the fibers from it also leave
through the depths. The neocortex is therefore doubled back on
itself, as it were, like the cerebellum. As in the case of the cerebellum, and the thalamus, we will depict the neuronal elements
in their actual relations.

The neuronal elements that are common to most areas of the
neocortex are illustrated in Fig. 61. The description is taken from
the accounts of Cajal (1911) and Lorente de Nó (1930), together
with more recent work of Garey (1971), Valverde (1971) and
Lund (1973) on visual cortex, and Szentágothai (1969), Jones
and Powell (1970a), and Schiebel and Schiebel (1970) in other
areas.

INPUT   The *afferent input* to the primary sensory cortex is carried, by definition, in the axons of the principal neurons of the

---

FIG. 61. Neuronal elements of human neocortex.

Inputs: specific sensory afferents (SA); nonspecific sensory afferents
(NSA).

Principal neurons: superficial pyramidal neurons (SP); deep pyramidal neurons (DP); giant pyramidal neurons (Betz cells) of the motor
cortex (GP); recurrent axon collaterals (rc).

Intrinsic neurons: granule cell (G); basket cell (B); bipolar cell
(BP).

Histological layers from I to VI are shown at left, together with a
diagrammatic representation of cell types and densities in the layers.

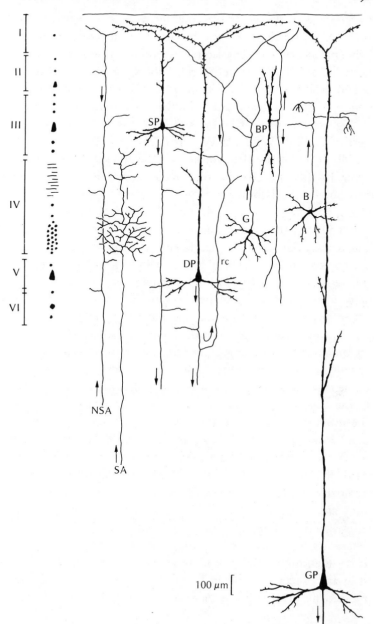

100 μm

thalamic relay nuclei. If we consider the visual cortex for purposes of illustration, the specific afferent fibers arise from the principal neurons of the lateral geniculate nucleus, which have been described in Chapter 9. The fibers are several microns in diameter and myelinated. They enter through the cortical depths and terminate with an extensive and dense arborization. As shown in Fig. 61, the field of arborization extends 100-400 μm in both vertical and horizontal planes. It is so placed as to be about in the middle of the cortical thickness, in a layer usually designated as layer IV (see Fig. 61); a few branches are often described as reaching layer III above.

In addition to this specific input, there is a variety of other types. One of the most common is shown in Fig. 61, a somewhat thicker, myelinated fiber that ascends through the cortex, giving off single branches at several levels, including the most superficial. There is a great variety in the branching patterns of these fibers, some fibers giving off branches to many layers (as in Fig. 61) and others giving off branches to specific layers only. These fibers have many origins. Some arise from the so-called *nonspecific thalamic nuclei,* carrying input from the *reticular-activating system* of the brain stem. Some arise directly from *brain-stem nuclei.* Some arise from the group of nuclei within the cerebrum known as the *basal ganglia.* And finally, many arise from other parts of the *neocortex:* from nearby areas of cortex (*short association fibers*), from distant areas (*long association fibers*), and from areas in the contralateral hemisphere (*commissural fibers,* passing through the *corpus callosum*). No attempt will be made to characterize these inputs further, other than to note that, as a group, they are different in their modes of termination from the specific afferent fibers. They may be considered analogous to the *central inputs* we have identified in other cortical regions. As such, they carry the inputs of most interest as far as "cortical" functions of man are concerned, but, in line with the remarks already made, they are unfortunately the least accessible for experimentation.

PRINCIPAL NEURON   The output from the cortex is carried in a type of neuron called a *pyramidal cell*. This is the principal neuron of the neocortex. Like pyramidal cells of other cortical regions, it derives its name from the shape of the cell body, which is imparted by the characteristic apical and basal dendritic trunks. There are, in general, two groups of pyramidal cells. One group is located in layers II and III. There is a size continuum in these layers, with the smallest and most superficial (SP in Fig. 61) cells about 15 $\times$ 30 μm, the largest and deepest cells (DP) up to 25 $\times$ 50 μm. The other, less numerous, group is in layer V, where the cell bodies, as seen in visual cortex, may reach dimensions of about 40 $\times$ 80 μm.

The single long *apical dendrite* ascends vertically through the cortex to branch and terminate in the most superficial (molecular) layer (layer I). The length of this apical dendrite thus depends on the depth of the cell body from which it arises. As can be seen in Fig. 61, this varies in visual cortex from a minimum of perhaps 200 μm (cell body in layer II) to a maximum of approximately 1600 μm (cell body in layer V). The superficial branches reach a radius of 200-400 μm. The apical dendritic trunk ranges in diameter from about 5 to 20 μm. These are therefore among the thickest, as well as longest, dendrites in the nervous system. A prominent feature is the numerous *spines*, particularly on the most superficial branches; we will discuss these in greater detail later.

The pyramidal cell also gives off several *basal dendrites*. The trunks vary up to 6 to 8 μm in diameter. They branch somewhat sparingly to an extent of 200 to 400 μm from the cell body. They also bear spines but not as many as the apical dendrites.

The axon leaves the deep aspect of the cell body or a large dendrite. A prominent feature is the *collaterals* given off by the axon while it is still within the cortex. These are of two main types. Some are short *horizontal collaterals*, given off to various layers during the descent of the axon (as shown for the SP cell in Fig. 61). Others are *recurrent collaterals*, which ascend through the cortex, varying greatly in the level they reach, the number of branches, and the lateral extent of the branching. An example of a recurrent collateral with limited branching and lateral extent is shown for the DP cell of Fig. 61. According to the Schiebels (1970), the greatest lateral extent for the branching field of collaterals from a single axon is of the order of 3000 μm (3 mm). To what extent such collaterals are part of the strictly local circuits in the cortex, and to what extent they represent intra- and intercortical association pathways, is a matter of definition.

This description summarizes the main aspects of the principal neuron as it applies to its geometry within the cortex. But one of the problems with the neocortex is that there seem to be exceptions to any rule. For example, some pyramidal cells have axons that do not leave the cortex; these appear, therefore, to be intrinsic neurons rather than principal neurons. On the other hand, some fusiform cells of the deep layers have axons that leave the cortex and must, therefore, be included as principal neurons. It is best to recognize this overlap of cell types as one of the characteristics of the neocortex, rather than draw hard and fast lines that do not, in fact, accord with reality.

The axons of the principal neurons have many destinations, which fall into two main groups. Some axons remain at the cortical level and serve to connect to other parts of the cortex; they are *association fibers*. The shortest of them (up to a centimeter or so) connect to other parts of the same cortical region and, therefore, overlap the longer recurrent collaterals mentioned above. Association fibers of medium length (up to several centimeters) connect to neighboring cortical regions. The longest fibers, many centimeters long, connect to distant cortical regions. Some (commissural) fibers pass through the corpus callosum to connect to cortical regions of the opposite hemisphere.

The other group of axons carries the cortical output to the rest of the brain; these are termed *projection fibers*. The subcortical projection sites are many and include most of the sites that have already been mentioned as sending the afferents to the cortex: the *thalamic nuclei* (both specific and nonspecific), the basal ganglia, many parts of the *brain stem*, and, finally, the *spinal cord*.

INTRINSIC NEURONS   The cells whose axons remain entirely within the local region of cortex in which they arise are usually grouped under the general heading of *stellate cell*. There are several types.

In primary sensory cortex, intrinsic neurons are found in great numbers in layer IV. Their distribution in layer IV is obviously very similar to, though not exactly the same as, the arborizations of the primary afferent fibers, as can readily be seen in Fig. 61 (Gr). By virtue of the relatively small size of their cell bodies (10-20 μm diameter) compared with the pyramidal cells, they are often called *granule cells*. We note, as we have noted before, that this term does not imply any similarity of structure or function with the granule cells of the olfactory bulb, cerebellum, or dentate fascia. Their numbers and density give a characteristic appearance to histological sections of sensory cortex and have become, thereby, the basis for distinguishing between sensory cortex (granular) and nonsensory, or motor, cortex (agranular).

The granule cells are multipolar neurons, sending out several dendritic trunks at apparently random orientations. The trunks are several microns thick, and they ramify to a modest degree through a field of 200-400 μm. The dendrites are described as having a modest investiture of spines or none at all (Colonnier, 1968). The axon is thin (several microns in diameter) and thinly myelinated. It generally is oriented vertically and ascends in the cortex to ramify largely in layer III, among the pyramidal and stellate cells there (see Fig. 61). The vertical extent of the axon may reach 500 μm or so.

Another type of stellate cell is found scattered throughout the middle and upper layers. The cell body is 15-30 μm in diameter. Its dendritic arborization is similar to that described above, but its axon distributes mainly in a horizontal direction (B, Fig. 61). The field of distribution is of the order of 300-600 μm. The axonal branches of this type of cell end in a characteristic tuft that forms a nest around a pyramidal cell body. In this respect this type of cell resembles the basket cells of the cerebellum. This type is scarce or absent in visual cortex (Lund, 1973), but prominent in motor cortex (Marin-Padilla, 1972).

A third type of stellate cell has, in contrast, an exclusively vertical orientation, as illustrated (V) in Fig. 61. As can be seen, it is, in fact, a *bipolar cell*, with superficial and deep dendritic fields. Its axon bifurcates into ascending and descending branches, which presumably carry the output of this cell to the most superficial and deep layers of the cortex.

A final type is a *fusiform cell*, situated in the deepest layers, with a relatively large cell body and several dendritic trunks with irregular orientations. The destinations of the axons are various: lateral, superficial, and deep. As already noted, some of these cells may send their axons out of the cortex, thus acting in fact as principal neurons.

These are the main types of intrinsic neuron in the neocortex, but they by no means exhaust the list. Among the others may be mentioned the cells of the most superficial cortical layer (layer I). These cells (H in Fig. 61) have a markedly horizontal orientation, reflecting the plexus of horizontal processes (terminal branches of apical dendrites and of axons) within which they lie. In his first description of the cortex (Cajal, 1890), Cajal reported that some of these neurons have more than one axon, each arising from a separate part of the cell body or dendritic tree; he termed them "cells with double axon." He later retracted this (Cajal, 1911), stating that, on re-investigation, one of the axons could always be identified as a dendrite. The rule that a neuron, if it has an axon, has only one, is so ingrained in our thinking that it might be of interest to reconsider these cells from that point of view.

Visual cortex is among the thinnest of the cortical regions. Motor cortex is, by contrast, among the thickest of the cortical regions. There the group of deep pyramidal cells, in layer V, is particularly numerous. They may have apical dendrites with lengths, in the human cortex, of up to 4 mm (4000 μm); an example is shown to the right in Fig. 61. These are probably the longest dendrites in the central nervous system. The cell bodies that give rise to these apical dendrites are also among the largest in the central nervous system. The largest measure up to 60 $\times$

120 µm, and are called the *giant pyramidal cells of Betz*. For many years it was thought that these giant cells perform some ultimate and commanding role in control of movement, but it is now clear that these cells are situated in the parts of the motor cortex that project to the motoneurons that innervate the distal muscles of the leg (see Walshe, 1947). Thus, their great size bears no relation to their complexity of function; it probably reflects, rather, the great length of axon that must be maintained.

This is a useful illustration of the fact that in the vertebrate brain the size of a neuron has no direct relation to complexity of function. If anything, the relation appears to be inverse; this, at least, was Cajal's belief in ascribing delicacy of function to the small (intrinsic) cells of the cortex. This is in contrast to the widely held assumption that a dominant feature in the organization of invertebrate nervous systems is the presence of large "command" neurons. It appears that the vertebrates have rejected an autocracy and opted instead for a committee system!

If we take an overview, what seems to stand out most clearly is the great variety of cortical neuronal elements. There are many different inputs and outputs; a wide variety, and overlap, within the types of principal and intrinsic neuron populations; and a multiplicity of intrinsic interrelations. The variety in all these categories is much greater than that present in other regions; conversely, there is little of a stereotyped nature in the elements, as is obvious in many of the other regions. From this one can surmise that the cortex is concerned with a minimum of fixed or obligatory functions; it provides rather a neuronal substrate that is generalized, possibly flexible, and presumably open and utilizable in a maximum of different ways.

CORTICAL CYTOARCHITECTONICS

We have described the main features that distinguish sensory receiving areas of the cortex and set them apart from the primarily motor areas. These constitute only a small fraction of the cortex. Within the remaining domains are many gradations between these extremes; these take the form of differences in the pattern of the layers: their presence or absence, thickness, cell types, cell densities, etc. The analysis of these differences constitutes the study of *cytoarchitectonics*. The neurohistologists of the early part of the century set about this task with zeal, and, while their political brethren carved up the map of Europe, they created their own states and duchies within the cortical subcontinent. Examples of the resulting maps are to be found in textbooks of neuroanatomy. The dimensions of cytoarchitectonic areas are in the range of 3-30 mm.

To discuss the significance of the cytoarchitectonic subdivisions of the cortex would take us far afield; a useful introduction and critique has been provided by Sholl (1957). Here we may simply note, by way of comparison, that the cytoarchitecture is regarded as essentially uniform throughout the olfactory bulb and the cerebellum, but non-uniform in the ventral horn, retina, thalamus, olfactory cortex, and hippocampus. In the neocortex, subdivisions appear to result from permutations of the variety of inputs, outputs, and intrinsic connections that have been mentioned above; one of the functions of the neocortex seems to be the elaboration of these permutations to the greatest possible extent.

Within the sensory areas are two aspects of cytoarchitecture that have particular relevance for synaptic organization. One is the strong element of vertical orientation that cuts across the cortical layers. This is a reflection of the vertically oriented axons and apical dendrites (Fig. 61). The possibility that the cortex might be organized into vertical columns of interconnected cells that extend across all the layers was recognized by Lorente de Nó (1938), and it has taken on special significance in view of the physiological evidence for functional columns within the cortex; we will review the evidence later. The diameter of the anatomical columns and the physiological columns both fall within the range of 100-500 μm.

A particularly clear manifestation of columnar organization is to be found in the islands of neuropil within layer IV of the somatosensory area of the rat cortex. Lorente de Nó (1922) first described them and showed that they are regions containing the terminal arborizations of afferent fibers and the dendrites of cortical neurons. He termed them "glomeruli" in analogy with the glomeruli of the olfactory bulb (see Chapter 6). They have recently been studied in detail by Woolsey and van der Loos (1969), who have shown that they measure from 100-500 μm in diameter and 100 μm in depth (i.e., the thickness of layer IV). Woolsey and van der Loos have characterized such a region as a "multicellular cytoarchitectonic unit", and have introduced the term "barrel" for it, citing as support an illustration of a barrel from the earlier work of Breughel (fl. 1560). One need intend no disrespect for such august authority if one notes that insofar as this is a circumscribed region providing for synaptic interconnections of afferent fibers and cortical neurons, it probably falls within the definition of macroglomerulus as discussed in Chapter 8. There is, indeed, a close similarity to the glomerulus of the olfactory bulb, in terms of position at the site of afferent input, dimensions, and in its being a region of synaptic processes surrounded by cell bodies. There is evidence that a cortical glomerulus receives the input from one vibrissa in the rat's snout and that within this input are several different sensory submodalities.

CELL POPULATIONS

It is a commonplace to cite an estimate of 10 billion neurons contained within the human neocortex as evidence for the fantastic overgrowth and therewith unprecedented capabilities for complex functions of this region. Those who cite this figure invariably fail to recall or realize that the number of granule cells in the cerebellum is probably several times this number (i.e., 20-40 billion; see Chapter 8); hence, it is not the large population of neurons per se that accounts for the special capabilities of the neocortex. The complexity of synaptic interactions within the cortex, and with other brain regions, are as already indicated, the crucial factors.

There is of course a dramatic increase in the population of cortical neurons as one ascends the vertebrate series. Among mammals with a well-developed cortex, there are also large differences according to size. Associated with this are differences in the density of neuron packing, with, in general, a relatively high density in a small brain and a low density in a large brain. The density also varies in different parts of the brain, being higher in the visual area than the motor area, for example, by a factor of about 3:1. In man the over-all packing density of cortical cells is about 10/0.001 mm³; the density in the visual area is about 70/0.001 mm³. By comparison, the packing density of granule cells in the cerebellum is about 2400/0.001 mm³. These and other quantitative data have been summarized and discussed by Sholl (1957).

There is, thus far, little evidence regarding the populations of principal and intrinsic neurons. One can only estimate that the principal neurons are quite numerous, and the I : P ratio must, therefore, be relatively modest.

SYNAPTIC CONNECTIONS

The first systematic study of cortical synapses with the electron microscope was by Gray in 1959. He reported that cortical synapses fall into two main groups, one (type I) characterized by extensive areas of contact and asymmetrical membrane thickenings, the other (type II) by smaller areas of contact and symmetrical thickenings. As noted in Chapter 2 and in subsequent chapters, these two types have been found in many parts of the brain and have served as a useful tool in identifying the synaptic connections made by a given class of neuron.

Type I synapses are mostly axodendritic, from axon terminals onto dendritic spines; examples are shown in Fig. 62(A). Gray was the first to describe in detail the *dendritic spine*: an elongated knob-like protuberance, 2-4 μm in length and 1-2 μm in diameter. These spines typically contain a collection of flattened sacs, or cisternae, as shown in Fig. 62(A). This was given the name of

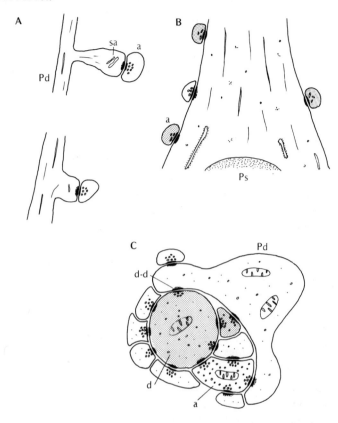

FIG. 62. Synaptic connections in the primate neocortex. (A) Axodendritic synapses from axon terminals (a) onto spines of pyramidal cell dendrites (Pd); note the spine apparatus (sa). (B) Axosomatic connections onto pyramidal cells. [(A)-(B) after Gray, 1959; Colonnier, 1968.] (C) Complex arrangements of synaptic connections in the primate motor cortex, including dendrodendritic connections (d-d). (After Sloper, 1971.)

spine apparatus, rather for want of a better term. For a time, it was assumed that a spine needed a spine apparatus to qualify as a spine, but it has developed that few of the spines in other parts of the brain contain organelles of this nature; since they are spine-like in appearance, they may also be referred to as spines, as we have done.

Type II synapses were reported by Gray to be preferentially located on the cell bodies of cortical neurons. Subsequent workers have incorporated this into the hypothesis of a preferential location of synaptic inhibition at near the site of impulse initiation in the cell body and/or axon hillock. We have had ample evidence in other regions, however, that the siting of inhibitory synapses is more complex than this, depending on the relations to excitatory inputs and to local dendritic input-output interactions as well.

A recent systematic study has provided evidence that the type I and II synapses within the cortex are extremes of a continuum that grades from the one into the other (Colonnier, 1968). In this study, the synapses onto pyramidal neurons were distinguished from those onto stellate neurons. Most synapses onto pyramidal neurons were onto the dendritic spines and fell into the category of type I. Synapses onto the pyramidal cell bodies were relatively few, and of type II. Stellate cells, on the other hand, were reported to have an intermingling of type I and II synapses onto both the dendrites and the cell bodies. For details of more recent work the reader is referred to Peters and Kaiserman-Abramof (1969), Jones and Powell (1970b), Garey (1971), and Le Vay (1973).

In the difficult task of sorting out the cortical neuropil, it has been commonly assumed that a presynaptic terminal is an axon and a postsynaptic terminal is a dendrite. That this assumption, based on classical notions, is now obsolete has been amply documented in the course of this book. The first evidence that this might also be true of the neocortex was provided by van der Loos (1959), who reported that, in Golgi-stained material, dendrites from two neighboring neurons could be seen crossing each other and making a synaptic contact. He called these dendrodendritic synapses, the first use of that term in the central nervous system. In the absence of electron-microscopic confirmation, however, this report remained unsubstantiated.

Very recently, electron-microscopic evidence for such connections has been obtained; an example from the work of Sloper (1971) is illustrated in Fig. 62(C). It can be seen that one of the dendrites (InD) has a synapse onto the spine of dendrite PD; InD is also presynaptic to another profile and postsynaptic to several terminals. In addition, numerous axodendritic synapses can be seen. This report awaits more extensive confirmation. Meanwhile, its significance rests in the demonstration of dendrodendritic synapses at the cortical level in the brain, and in a motor region.

Several other types of synaptic connection have been demonstrated in recent work. One of these is a connection from an axon collateral back onto the dendrite of the pyramidal neuron from which the axon arose; this has been termed an *autapse* by van der Loos and Glaser (1971). The evidence was obtained from Golgi-stained material and awaits confirmation with the electron microscope. There is the implication that the collateral might also connect to other pyramidal neurons. We have seen examples of synapses between two intrinsic neurons of the same type among the amacrine cells of the retina. We have also seen an example of this type of connection between output neurons among the Purkinje cells in the cerebellum (where the connections are inhibitory) and between pyramidal neurons in the olfactory cortex and hippocampus (where the connections are excitatory).

Among the other types of connection may be mentioned axo-axonic synapses onto initial segments arising from pyramidal cell bodies (Peters, Proskauer, and Kaiserman-Abramof, 1968); the similarity to the olfactory cortex has already been mentioned (Chapter 10). The prevalence and significance of this and other varieties of synapses awaits further study.

BASIC CIRCUIT

In most of the regions of the brain considered in this book, it has been possible to assimilate the information about neuronal elements and their main synaptic connections into a summary diagram. In those regions, the connections and pathways illustrated by the diagram are either obvious consequences of the internal architecture, or they represent a broad consensus drawn from mutually supportive evidence from many workers.

In considering the neocortex, we forsake those warm precincts. Relatively little insight into synaptic organization is provided by the disposition of the neuronal elements themselves, whereas this information by itself was sufficient for Cajal and his contemporaries to deduce the basic pathways in many other parts of the brain. The more recent studies have only begun to close the gap between the description of synaptic connections and the identification of the neuronal elements from which those connections arise. Before a consensus on these matters is possible, there must be answers to such basic questions as what connections are made by the input fibers onto principal and/or intrinsic neurons, and what connections are made between principal and intrinsic neurons. To complicate matters, it seems clear that these connections will differ in kind as well as degree in the different regions of the cortex.

In the face of such problems, a diagram of cortical organization is less a summary of present knowledge than a guide to the important questions that need to be answered. The diagram of Fig. 63 is intended to be such a guide.

To begin with the *specific afferent fibers* (SA), let us note the possibilities: they could provide input to the principal neurons, the intrinsic neurons, or both. What is the evidence?

With regard to the principal neurons, the fact that the specific afferent fibers terminate in layer IV means that they can connect only to the apical dendritic trunks of the deep pyramidal neurons (DP), as indicated in Fig. 63. Now, deep pyramidal neurons are, in fact, relatively scarce in the sensory cortex; such connections, however, might be more important in the motor cortex, in which the population of deep pyramidal neurons in layer V is numerous. With regard to the intrinsic neurons, their localization in layer IV accords well with the level of termination of the specific afferent fibers. In addition to this localization, there are many intrinsic neurons in the sensory regions to which the afferent fibers project. This interpretation of Golgi-stained material is supported by the experimental studies of synaptic connections mentioned in the previous section, and it is also

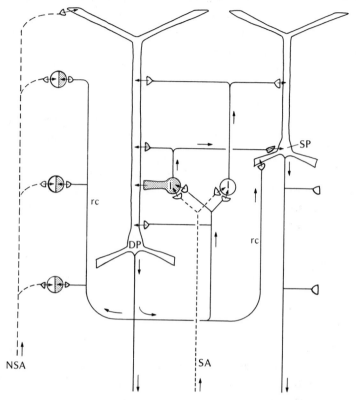

FIG. 63. Basic circuit diagram for the neocortex. Abbreviations as in Figs. 61 and 62; the intrinsic neurons are summarized together under (I).

supported by the physiological studies of synaptic actions to be described in the next section.

The diagram of Fig. 63 illustrates both these possible connections of the afferent input. Note that the input fibers, together with the principal and intrinsic neurons, form our familiar triad of synaptic elements. The relative proportions of the elements of the triad are apparently different in motor and sensory areas of the cortex. It seems particularly significant that the input to sensory areas is directed mainly to the intrinsic neurons, because it implies that the intrinsic neurons act, in effect, as an additional relay along the input pathway. The presence of a large population of intrinsic neurons, the function of which is to relay inputs to the principal neuron of a cortical region, is, of course, reminiscent of the population of granule cells with a similar function in the cerebellum (Chapter 8). It appears that, in addition to the thalamic relay, there is an additional staging or processing station within the cortex itself. One may also note the comparison with the dentate fascia relay to the hippocampus (Chapter 11).

Cajal noted the large populations of small neurons in the cortex and suggested that they were correlated with delicacy of function. Our study of synaptic organization indicates that this delicacy must be assessed in the light of whether the functional position of the neurons is one of vertical relays along the input pathways or horizontal processing through cortical circuits. It appears that the former is the case in the sensory regions of the cortex.

Pathways for nonspecific afferent fibers (NSA) are indicated by a dotted line to the left in Fig. 63; possible inputs are shown to either principal or intrinsic neurons, with either excitatory or inhibitory synaptic actions, at all levels of the cortex. These kinds of afferents predominate in the motor areas and in the association areas of the cortex; the latter areas are, as already noted, correlated with the evolutionary rise of human intelligence. Obviously, there is much work ahead in obtaining evidence for the synaptic basis of the quintessentially human aspects of behavior mediated by these afferents and areas.

Let us next turn to *intracortical pathways*. The most prominent pathways for internal connections appear to pass through the axon collaterals of the pyramidal neurons. The possible connections that these collaterals might make within the cortex are illustrated in Fig. 63. It can be seen that, by and large, these connections involve the same possible sites as do the external input axons. There seems to be much more overlap in this regard than in other regions of the brain; in the hippocampus, for example, the axon collaterals have connections that are quite distinct from those made by the input fibers (Chapter 11). Whether the internal circuits feed back onto the same sites of external input, or onto distinctly different sites, is an important point for understanding the dynamics of synaptic interactions in the cortex.

With regard to the sequences of activity within intracortical pathways, let us first consider possible excitatory actions. If attention is directed to the recurrent collaterals in Fig. 63, they can be seen to provide possible excitatory inputs from one principal neuron to another (DP→SP or SP→DP) as well as to intrinsic neurons (In). Such connections provide for rapid dispersion of activity throughout a population of neurons, a fact Cajal (1911) recognized and termed *avalanche conduction*. It can readily be imagined that neurons thus activated could excite other neurons, and so on. The possibility of such sequences was recognized by Lorente de Nó (1938), and they were conceptualized as *multiple chains* of neurons, with some of them providing for *self-re-exciting* or *reverberating circuits*. This notion, like so many others, has a long history, deriving, in part, from the simple model of circus movement for fibrillation in the heart. It is widely quoted in textbooks, but in the cortex, as elsewhere, it has remained very much a hypothesis (see Mountcastle and Darien-Smith, 1968). We will examine the evidence for sequences of action within cortical circuits in the next section.

Lorente de Nó (1938) further developed the concept that the chains of neurons within the cortex are organized into *vertical cylinders;* in his words,

> All the elements of the cortex are represented in it, and therefore it may be called an *elementary unit*, in which, theoretically, the whole process of transmission of impulses from the afferent fiber to the efferent axon may be accomplished.

We will see that recent physiological studies support this concept as it relates to certain aspects of the incoming sensory input and its initial stages of processing within the cortex. That the cylinders or columns are organized in relation to the arborization of afferent terminals was inferred by Lorente de Nó, and it is supported by the evidence that has been obtained for cortical macroglomeruli (barrels) (see above). Whereas Lorente de Nó conceived of the activity within the vertical chains as involving sequences of excitation, it seems likely that it involves complex excitatory and inhibitory interactions.

The neuronal chains were originally conceived of as involving mainly internuncial (intrinsic) neurons, but, in view of the re-excitatory circuits through principal neurons onto other principal neurons in the olfactory cortex (Chapter 10) and hippocampus (Chapter 11), it seems likely that such circuits are present in the neocortex as well. The evidence for autapses among the synaptic connections (see previous section) may be recalled in this regard. It may be noted that the organization of such circuits into columns can only occur in a cortex in which there is vertical dispersion of the pyramidal neurons. There is limited vertical dispersion in the olfactory cortex and none in the hippocampus, so the re-excitatory chains in those regions have, perforce, a lateral rather than columnar organization.

Recurrent collaterals may be part of inhibitory as well as excitatory circuits, and in motor cortex there is ample evidence of recurrent inhibition of the pyramidal neurons, as will be described in the next section. It has been inferred that this is mediated by an axon collateral-interneuron pathway, in analogy with the Renshaw pathway in the spinal cord (Chapter 5). In the diagram of Fig. 63, this pathway passes through the elements labeled rc and I. Since basket cells are prominent in motor cortex, it has been conjectured that they are the inhibitory interneurons in this recurrent pathway. The interplay of feedback and feedforward inhibition with the re-excitatory circuits that have just been described probably is the basis for much of the complex processing that takes place within the cortex (see below).

The fact that cortical connections are made within a framework of horizontal and vertical pathways is indicated in Fig. 63. The *horizontal* connections have various extents, reflecting such factors as the fields of the axon collaterals and dendritic trees, the barrels and columns, the association fibers, and various other arch-itectonic characteristics. All of the factors that can enter into the horizontal parcellation of a region—different input sources, dif-ferent intrinsic circuits, different output destinations—are in-volved in the parcellation of the neocortex and would have to be included in the basic circuit of a given cortical region. We have noted that these factors stand out in the hippocampus, but the neocortex is even more complex in this region; the complexity of parcellation may, in fact, be regarded as one of the distinguishing characteristics of neocortical organization.

In the *vertical* dimension, the lamination of the cortex has been traditionally considered one of its hallmarks, but in terms of synaptic organization it can be seen that there is a great deal of overlap between the elements of different levels. Pyramidal cell bodies, for example, are found at several levels, in contrast to the thin layer that is characteristic of the principal neurons in the other laminated regions we have considered (olfactory bulb, retina, cerebellum, hippocampus). The axon collaterals of the pyramidal neurons terminate at all levels of the cortex, in con-trast to the more restricted terminal sites in most of the other regions. Similar remarks can be made about the nonspecific affer-ent fibers. Only the specific afferents to the primary sensory areas of the cortex seem to be as confined in the vertical dimension as their counterparts in the other regions.

The neocortex can thus be seen to be rather generalized in its organization in the vertical dimension, less severely laminated at the level of synaptic organization than is implied by the classical studies of cortical architectonics. This is not to deny that the cortex is a cortex, in view of the characteristics of synaptic organization that have been discussed above and in Chapter 10. The great overlap of neuronal elements and synaptic connec-tions, however, does give the impression of a generalized organ-izational matrix more akin in some respects to that of the ventral horn of the spinal cord (Chapter 5) than of such severely lami-nated structures as the olfactory bulb, retina, and cerebellum. Perhaps this, in some way, reflects the position of the neocortex as a penultimate and pre-motor station for many pathways to the motoneurons of the brain stem and spinal cord.

SYNAPTIC ACTIONS

The considerable overlap of neuronal elements and synaptic connections within the neocortex has presented electrophysiologists with problems not unlike those that arise in studies of the spinal cord, but without the redeeming feature of separate input and output pathways that have been the basis for the analysis of spinal organization (Chapter 5). There is in the neocortex one discrete pathway, however, and that is the *corticospinal tract*, an output pathway that arises from the pyramidal neurons of the motor cortex and descends unbroken to the spinal cord. It is represented in Fig. 1. It is particularly prominent in primates; as described in Chapter 5, it provides an important input to the motoneurons that control the hand (Phillips, 1971). Because the fibers are gathered together in a pyramid in the brain stem, the pathway is called the *pyramidal tract*, and, of course, this has led to confusion over whether a neuron is a pyramidal neuron because of its shape or because it sends its axon through the pyramid. The reader should be clear in his own mind that only a small proportion of pyramidal-shaped neurons are also pyramidal tract neurons, whereas most of the latter are the former.

The importance of this tract to electrophysiologists is that it enables them to identify a pyramidal cell from which they are recording by the ability to backfire it (antidromically) through its axon; this is the same method used to identify motoneurons and the principal neurons in other parts of the brain. The first intracellular recordings in the brain, following the pioneering work on the spinal cord, were obtained from cortical neurons identified in this way (Phillips, 1956).

AFFERENT EXCITATORY AND INHIBITORY ACTIONS In analyzing synaptic actions on cortical cells, we first ask questions that concern the site of afferent input and the nature of that input; these

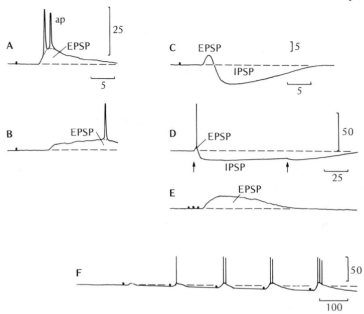

FIG. 64. Synaptic actions in the neocortex. (A) Intracellular response of a cortical neuron to a volley in the deep radial nerve. (B) Response to a volley in the superficial radial nerve. (C) Response to a volley in the dorsal column of the spinal cord. [(A)-(C) after Oscarsson et al., 1966.] (D) Prolonged antidromic stimulation (between arrows) showing an EPSP-IPSP sequence. (After Phillips, 1959.) (E) EPSP response to antidromic volleys (Phillips). (F) Intracellular response to slowly repeated volleys from the nonspecific thalamic nuclei. (After Purpura et al., 1964.)

questions parallel those asked by anatomists, as previously discussed. This has been investigated by recording the responses of cortical neurons to volleys in the sensory input pathways (Towe, Patton, and Kennedy, 1964; Amassian and Weiner, 1966; Watanabe, Konishi and Creutzfeldt, 1966; Oscarsson, Rosén, and Sulg, 1966; Swett and Bourassa, 1967). Representative results are illustrated in Fig. 64(A), for the case of a volley set up by a single shock to a peripheral nerve that carries fibers mainly from muscle spindles in the arm (deep radial nerve). The response consists of a short-latency EPSP, which leads to the discharge of one or two impulses. Note the brief duration of the EPSP (similar to that in a motoneuron).

The short latency has been interpreted as indicating that the pathway from the nerve to the cortex has a minimum of synaptic relays; as can be seen by tracing the connections in Fig. 1, this means three relays, in the dorsal column nuclei, the thalamus, and, finally, the cortex. This means, in particular, that there is a monosynaptic input from the thalamocortical relay axon to the cortical neuron. By comparison, a volley in a peripheral nerve that contains mostly fibers from the skin (superficial radial nerve) gives rise to a response, in the same cortical neuron, which consists of a long-latency EPSP with a much slower time course [Fig. 64(B)]. This demonstrates that there is at least one additional synaptic relay at the cortical level—a disynaptic or polysynaptic input, in the terminology used for the spinal cord.

The monosynaptic responses are located preferentially in the middle cortical layers where the afferent fibers terminate. What is the identity of the cortical neurons giving these responses? One study reports that they are pyramidal tract neurons (Swett and Bourassa, 1967); another reports that they are non-pyramidal tract neurons (Oscarsson et al., 1966). This point, therefore, remains in doubt.

Inhibitory actions are also present in the responses of cortical neurons to a peripheral volley; an example is shown in Fig. 64(C). It can be seen that the response consists of an initial depolarization followed by a long-lasting hyperpolarization. There is an obvious similarity to the excitation-inhibition sequences that we have seen in the responses of principal neurons in many parts of the brain to a volley in an input pathway (see Purpura, 1967).

In summary, these and other studies have shown that a cortical neuron may receive both monosynaptic and polysynaptic inputs from the periphery and that there is a convergence of different modalities onto the same neuron. They have shown that there are different patterns of spatial convergence, some neurons having wide peripheral fields and others having very small fields. There are different excitatory-inhibitory sequences. Some cortical neurons give rapid (phasic) responses, others give slow, maintained (tonic) responses. These synaptic actions provide some of the basis for the complex operations the cortex performs in its input-output relations (see below).

RECURRENT EXCITATORY AND INHIBITORY ACTIONS   Synaptic actions in the cortex are also revealed by the antidromic volley in the pyramidal-tract axons that invades collaterals within the cortex. A typical result (Phillips, 1959; Stefanis and Jasper, 1964) is shown in Fig. 64(D). The response consists of the antidromic spike invading the neuron, followed by a long-lasting hyperpolarization. The hyperpolarization can be shown to be due, not to the spike afterpotential, but to an IPSP. The latency of onset is sufficient for at least one synaptic relay, and it has been postulated that there is a circuit from axon collaterals through an inhibitory neuron back onto the pyramidal cells, in analogy with the Renshaw pathway in the spinal cord. This circuit is shown in the diagram of Fig. 63.

The inhibitory neurons in this postulated pathway may be basket cells, but the evidence is not conclusive; few cells can be found that discharge a train of impulses as do Renshaw cells in the spinal cord (Brooks, 1967). This might indicate that the inhibition is mediated by graded synaptic action, as though a dendrodendritic pathway from the intrinsic neurons onto pyramidal neurons, in analogy with retinal neurons (Chapter 7), olfactory granule cells (Chapter 6), and, possibly, the short-axon cells of the thalamus (Chapter 9). The finding of dendrodendritic synapses in the neocortex, as described earlier in this chapter, suggests that this possibility deserves serious consideration.

In addition to recurrent inhibition, depolarizing-hyperpolarizing sequences are also evoked through stimulation of the pyramids, as shown in Fig. 64(E). Although spread of excitation to include afferents from the brain stem to the cortex might be a factor in some experiments, these and other results have indicated the likelihood that pyramidal neurons excite other pyramidal neurons through a direct connection by their recurrent collaterals.

This is, therefore, evidence for the re-excitation loops previously discussed. It appears that these actions are mainly directed from smaller, superficial neurons to larger, deep ones, providing a possible mechanism for rapid enhancement of phasic drives by the largest output neurons to the motoneurons (Oshima, 1969; Purpura, 1972).

SENSORY STIMULI The experiments just described have utilized electrophysiological methods for the study of synaptic actions. A different approach has been to analyze responses to natural stimulation of peripheral receptors in the sensory areas of the cortex. This approach is not so effective for direct study of synaptic connections and synaptic actions, but it is obviously much more relevant to the natural functioning of the cortex. Its effectiveness for this purpose has depended on the ability to very strictly control the stimulus parameters. In the temporal domain, this has meant the use of pulses and steps of sensory stimuli; in the spatial domain, it has meant the use of stimulus spots, points, annuli, edges, arcs, etc.

The application of these methods, in conjunction with microelectrode recordings, to the analysis of intracortical organization was introduced by Mountcastle (1957) in his studies of somatosensory cortex, which receives its input from peripheral receptors via the dorsal column nuclei (Fig. 1) and the ventrobasal complex (Chapter 9). We will follow Mountcastle and Darien-Smith's (1967) account in summarizing the essence of the results.

When microelectrode penetrations are made vertically through the cortex, all the neurons encountered are of the same modality and approximately the same peripheral receptive field. The cells of the middle layers (III and IV) are activated earliest by an afferent input; those of the deeper layers have longer latencies. Thus, within a few milliseconds, allowing for only a few synaptic relays within the cortex, the activity arriving through the input is "translated", to use Mountcastle's phrase, or processed, vertically through the superficial pyramidal neurons to yield an output through the deep pyramidal neurons. From these findings, it is deduced that the *elementary functional unit* of the cortex is a vertically oriented column of cells that composes an input-output linkage. The physiological evidence thus supports and extends the inferences of Lorente de Nó drawn from the vertical orientation of the neuronal elements (see above). The columns for different modalities are sharply demarcated from each other, and there is evidence (e.g., Mountcastle and Powell, 1959) that activation of one column produces a surround inhibition of neurons in neighboring columns.

The organization into columns has been found in visual and auditory as well as somatosensory cortex. The results in visual cortex have been especially clear, because of the nature of the stimulus control and the fact that the sequence of intracortical processing has been followed through several integrative stages. As is well known, this work is associated with the names of Hubel and Wiesel (1962, 1968). We may recall that in the simple retina of the primate (Chapter 7) the ganglion cells are organized in terms of centers and surrounds of light, and this holds also for the relay through the lateral geniculate nucleus (Chapter 9). In the visual cortex, however, to which the lateral geniculate nucleus projects, the simplest responses are to bars or edges of light, shone on a part of the retina with a particular orientation; the cells giving these responses are termed *simple cells* [Fig. 65(A)]. More complex responses are to a bar or an edge moving over the retina in a particular direction; the cells giving these responses are termed *complex cells*. Typical responses of a cell of this type are illustrated in Fig. 65(C). The most complex responses (by *hypercomplex cells*) are to bars or edges, moving with particular orientations, with various critical dimensions, and with antagonistic regions within their peripheral fields.

From these responses, Hubel and Wiesel have deduced that there is a sequence of processing within the cortex, through the simple cells to the highest orders of hypercomplex cells and beyond. The sequence takes place within a column of cells, as shown by the fact that all the cells encountered in a vertical microelectrode penetration have the same orientation to their

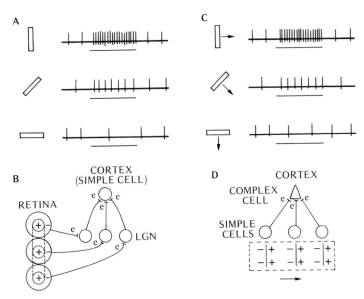

FIG. 65. Responses of neurons in the visual cortex to stimulation of the retina with light. (A) Extracellular recordings of unit spikes. Responses of a "simple cell" are shown to stimulation of the retina with a bar of light with three orientations; time of stimulation shown by the horizontal bar beneath the recording trace. (B) Schematic diagram to illustrate possible synaptic connections mediating the responses in (A). (C) Responses of a "complex cell" to a bar of light, moving as indicated by arrows. (D) Schematic diagram to illustrate possible synaptic connections mediating the responses in (C). (After Hubel and Wiesel, 1962; Michael, 1969.)

receptive fields. The sequence from the retina to simple cells is illustrated in Fig. 65(B), in terms of the consequent synaptic connections. The convergence of simple cells onto complex cells is illustrated in Fig. 65(D). These are formal diagrams of synaptic connections and need to be substantiated experimentally. Recent studies of neurons identified with intracellular stains have shown that simple cells are mostly (although not exclusively) stellate neurons in layer IV and that complex and hypercomplex cells are mostly pyramidal neurons in the more superficial and deeper layers (van Essen and Kelly, 1973). This is consistent with the sequence of monosynaptic and polysynaptic responses deduced from the electrophysiological experiments described above.

The sequence within a column is presumably mediated by vertical circuits providing for excitation, re-excitation, and excitatory-inhibitory interactions, as discussed previously in relation to the diagram of Fig. 63. The important point about these sequences is that, rather than providing for a wider dissemination of the input, they provide for a narrowing of responsiveness, a finer tuning, to more specific aspects of the input. Thus, whereas a retinal ganglion cell responds to diffuse light falling on the entire retina, a hypercomplex cell in the cortex responds not at all to this nonspecific stimulus, but discharges vigorously when a specific pattern falls in a specific part of the retina with a specific orientation and rate of movement. This is thought to be the basis for the abstraction and generalization of patterns in the visual world.

MOTOR CORTEX The investigation of cortical motor function has one of the longest traditions in the history of brain physiology, originating in the studies of movements elicited by electrical stimulation of the cortex by Fritsch and Hitzig in 1870. The movements take the form of simple muscle twitches, from which it has been inferred that, during normal functioning, populations of cortical cells in the area stimulated control the spinal motoneurons that, in turn, innervate those muscles. In the intervening years, the classification of the populations of cortical cells has been increasingly refined. Specific *colonies* that control flexor and extensor motoneurons have been identified. Some of these colonies are narrow, others are wide (Landgren, Porter, and Phillips, 1962). Evidence has been adduced that a narrow colony has a *columnar* organization, in analogy with the columns of sensory cortex (Asanuma and Rosén, 1973). These aspects of organization have been demonstrated for the pyramidal tract neurons of the motor area, but it is well to point out that many other areas of cortex also contribute to motor behavior.

In the analysis of cortical organization relative to motor control, an important development has been the ability to record the activity of pyramidal tract neurons in an awake animal as it performs specific tasks. This approach, in the awake monkey, has been pioneered by Evarts (1969). These studies have begun to provide information about the timing of impulse output in relation to the initiation and the maintenance of a movement or a posture. One of the most relevant studies for the synaptic organization of the cortex has been addressed to the problem of whether the impulse output of a cortical neuron is correlated with the distance over which a particular movement is carried out, or the force necessary to move over that distance. The results have shown (Evarts, 1969) that there is a high correlation between impulse frequency and the force exerted to move a limb. These experiments indicate that there is not a simple mapping from impulse frequency to linear displacement.

These results have carried the implication that feedback from the muscle spindles is a necessary input to the cortex in order to keep it informed of the progress of a movement under different conditions of loading. From this it has been inferred that there must be some type of servo-loop through the cortex, superimposed on the reflex arc or loop at the segmental level in the spinal cord. The relation between the two loops can be seen in the diagram of Fig. 1. Phillips (1969) has shown that there is, in fact, a correlation in the synaptic organization at the two levels, in that the spinal motoneurons that innervate the hand are those that not only receive the most monosynaptic (Ia) input from the muscle spindles, but also the most monosynaptic input through the corticospinal fibers from the motor cortex (see Chapter 5). In his words,

> It is almost as if these particular spinal motoneurons had been transplanted into the cerebral cortex: as if the dendrites of the corticospinal pyramids were acting vicariously for the dendrites of the alpha motoneurons as antennae sensitive to intracortical synaptic activities.

To actually specify those activities, and the servo-loops or other pathways through which they are mediated, will be a difficult task (see Wiesendanger, 1973). An intriguing aspect of the problem is its relation to the question of consciousness. Merton (1964) has shown that we are not conscious of input from the muscle spindles when a muscle is moved passively, even though the spindle input does reach the level of the cortex. During willed, voluntary movement, however, the sense of effort that we are aware of does seem to arise from the spindle input. It may be possible, therefore, that what is termed awareness or consciousness in this particular experimental situation can be correlated with the synaptic pathways for the spindle modality in the cortex.

RHYTHMIC POTENTIALS The synaptic actions that have been described thus far have been produced by single electrical shocks or by discrete natural stimuli; the responses have taken the form of relatively simple synaptic potentials and impulse discharges that have their counterparts in the spinal cord and in other brain regions. But cortical responses are by no means all of this stereotyped nature. For example, stimulation of a nonspecific thalamic nucleus produces the responses shown in Fig. 64(F). The response consists of an EPSP that has a long latency, a long duration, and a low amplitude; in the first response of Fig. 64(F), the depolarization is below threshold for impulse initiation. This has been characterized as a *conditional* type of response, in contrast to the powerful *unconditional* response to specific thalamic nuclei (Purpura, Schofer, and Musgrave, 1964). We have noted in Chapter 11 that the responses of hippocampal pyramidal cells can also be characterized by their conditional and unconditional nature.

During *repetitive stimulation* the response in Fig. 64(F) changes, building up from a subthreshold depolarization to one that triggers three impulses; the EPSP's are superimposed on a gradually increasing hyperpolarization. The buildup in the responses bears some resemblance to that elicited by repetitive stimulation of the hippocampus and dentate fascia (Chapter 11), although it is less dramatic. As in the latter cases, some of the buildup may be due to an increased potency of the synapses. In the cortex, however, much of this effect is believed to be due to spread of activity within the circuits of the thalamus itself (see Purpura, 1972). This reflects the fact that the cortex and the thalamus are closely related, and it is a reminder that cortical function must be viewed within the context of corticothalamic organization.

As is well known, the *resting activity* of the cortex, recorded by electrodes on the scalp, takes the form of potential waves occurring at 8-12/sec, the so-called *alpha rhythm*. The rhythm is somewhat faster than that of the theta waves of the hippocampus (Chapter 11). We have discussed in Chapter 9 the postulated mechanism whereby the cortex is driven by input from the thalamus; the activity of the thalamus is rhythmically generated by a combination of membrane excitability changes and an inhibitory phasing of impulse discharge through recurrent feedback circuits. Similar mechanisms within the cortex itself may also contribute to the rhythmicity of the cortex. The potentials recorded from the scalp are due to summed currents generated by the entire ensemble of active neurons. To discuss this complex subject would take us far afield. For an extensive discussion of the mechanisms of the alpha rhythm, the reader is referred to Andersen and Andersson (1968).

We have discussed the tendency of the hippocampus to develop *seizure activity* (Chapter 11), and the neocortex shares in this property. One form of activity consists of large depolarizing waves that generate a burst of impulses, the *paroxysmal depolarizing shift*. As in the case of the hippocampus, the factors that contribute to the development of such activity include altered excitability of the neuronal membrane and abnormal balance of activity in synaptic circuits. The re-excitatory axon collaterals in cortical regions would appear to be a potent source of excessive excitatory drive that is absent in most of the other regions we have considered. Another factor is the accumulation of ions in the intercellular spaces; we have mentioned some of the effects this may have on neuronal excitability in Chapter 3, and there is current interest in the possible role of extracellular K in producing long-lasting depolarizations of both neurons and glial cells in the cortex. These and other aspects of synaptic organization as they relate to the mechanisms of epileptic seizures are discussed by Ayala et al. (1973) and Jasper, Ward, and Pope (1969).

DENDRITIC PROPERTIES

In our studies of other regions of the brain, we have seen that an understanding of synaptic organization requires knowledge of dendritic properties. This is particularly true for the neocortex, in which the apical dendrites of the pyramidal neurons constitute the main vertical elements of the cortical framework and the major substrate for synaptic connections. Cortical dendrites may have presynaptic as well as postsynaptic positions, as noted previously; the evidence for the former is still tentative, however, and attention may be confined for present interests to the postsynaptic functions. We first consider the electrical parameters, then the transfer through the apical dendritic trunk, and, finally, local dendritic properties with special reference to dendritic spines.

ELECTRICAL PARAMETERS   The whole neuron resistance (R_N) of a pyramidal neuron to the flow of current injected through an intracellular electrode in the cell body has been found to lie in the range of 4-15 MΩ, with an average of about 8 MΩ (Takahashi, 1965; Lux and Pollen, 1966). The charging time constant (τ_N) has been found to be approximately 8 msec. Both these values are similar to those for hippocampal neurons (Chapter 11) and several times greater than those for spinal motoneurons (Chapter 5).

To obtain values for the specific membrane resistance (R_m) an estimate of dendritic dominance (ρ) is needed (cf. Chapter 4). Under assumptions of a range for this factor of from 3 to 10, a range of 1500-4000 Ω cm^2 was obtained (Lux and Pollen, 1966). This is somewhat higher than the estimates for the spinal motoneuron. The higher resistance and the longer time constant enhance the spread of synaptic current through the dendritic tree of the pyramidal neuron. Similar observations have been made with regard to the hippocampal pyramidal neurons (Chapter 11).

In using these figures to estimate electrotonic lengths of apical dendrites, we must take into account the fact that the dendrites vary in both diameter and length, and therein emerges an interesting fact. We have discussed in Chapter 4 the point that the characteristic length (λ) varies with the square root of the diameter (d), with the consequence that electrotonic spread is relatively effective even in thin dendrites. Thus, even though the larger dendrite provides for more effective current flow over a given distance, the longer length produces more severe attenuation over the whole extent. If we compare the diameter and length of the apical dendrites of small pyramidal cells in the superficial layers of the center and large cells in the deep layers, it appears that not only does the scaling principle (introduced in Chapter 5) apply but that it may favor the smaller ones.

SYNAPTIC TRANSFER AND LAMINATION   We are now in a position to discuss the sites of different synaptic inputs to the pyramidal dendritic tree. The locus of generation of a synaptic potential has been assessed, using the same criteria as used for neurons of other regions: the relative amplitude of the synaptic potential, its shape index (see Chapter 4), its sensitivity to currents and to ions injected into the cell body. Many of these studies have been discussed by Purpura (1972), and Fig. 66 provides a schematic summary, similar to that presented for the spinal motoneuron (Fig. 23).

The essential results are, first, that monosynaptic and disynaptic inputs from the specific thalamic relay nuclei (see typical responses in Fig. 64) are sited proximally on the dendritic tree and cell body. The excitatory inputs are, thus, in a position to trigger impulses in the axon hillock; the inhibitory inputs are in an optimal position to oppose impulse generation; and the overlap in the distribution of the two types of input provides for maximum dynamic interaction between them. Presumably, there is a complex summing of many individual inputs that leads up to the

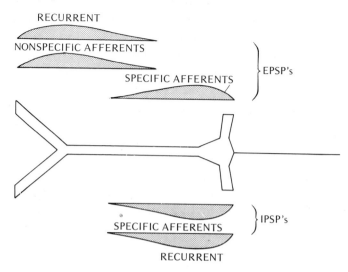

RECURRENT

NONSPECIFIC AFFERENTS

SPECIFIC AFFERENTS

} EPSP's

} IPSP's

SPECIFIC AFFERENTS

RECURRENT

FIG. 66. Distribution of some synaptic inputs to the pyramidal cell of the neocortex.

initiation and control of impulse generation. In these aspects, synaptic integration in the cortical pyramidal cell is similar to the spinal motoneuron (see Chapter 5).

The cortical pyramidal cell also has inputs preferentially sited on the distal branches of the apical dendrite; among the inputs that have thus far been identified are excitatory synaptic actions by the nonspecific thalamic nuclei and the re-excitatory collaterals of the pyramidal cell axons (see Fig. 66). The question then arises: How can these distal inputs have any effect on the integrative activity of the pyramidal cell?

In order to answer this question, calculations of the electrotonic properties of the apical dendrites have been made by Jacobson and Pollen (1968) using the methods of Rall. Starting with the values of whole neuron resistance previously obtained (see above), a study was made of Golgi-stained neurons in the cat neocortex in order to determine first the contribution of the apical as distinct from the basal part of the dendritic tree. The first major branch point in the apical shaft was found to be about 250 μm from the cell body. It was calculated that a steady synaptic potential at that point would decrement to 25-40% in spreading to the cell body. A depolarization of 20 mV throughout the dendritic tree distal to that point would therefore produce 5-8 mV of depolarization at the cell body; such depolarizations are, in fact, produced by strong stimulation of the nonspecific afferents from the thalamus, which are believed to have the majority of their terminals onto the superficial dendrites (see Fig. 66).

What then of inputs to the most distal branches? It was calculated that synaptic potentials in apical dendrites 750 μm from the cell body would decrement to only 2-3%; thus, a 20-mV depolarization at that distance would produce less than 1 mV of change in the cell body. Although this is only a fraction of the total depolarization needed to raise the membrane potential to the threshold for impulse discharge (5-10 mV), studies have shown that such small depolarizations can have marked effects on impulse frequency when the neuron is near the threshold level (Nacimiento, Lux, and Creutzfeldt, 1964). Thus, a modulating role of distal dendrites could consist of a refinement in the firing rates of neurons already brought close to firing level by the afferent systems that terminate closer to the cell body (Jacobson and Pollen, 1968).

We may also note that, with respect to transfer through the dendritic tree, the apical dendrite can be viewed as a series of loci for local integration, each locus having its maximum effect on its immediate neighbors. By this means, a given locus can bias or modulate the effect of an input to neighboring loci, an effect that takes on added significance if it is of a long-lasting or plastic nature such as might underlie learning or memory (see below). A dendritic site can also have a local output function through a dendrodendritic synapse; it remains to be determined how important such functions are in the cerebral cortex, in comparison with the olfactory bulb, retina, and thalamus.

In the course of this book we have seen that there are various mechanisms by which the transfer of signals through a dendritic tree can be enhanced. One is a *higher membrane resistance*, which provides for increased passive spread of electrical current; we have already noted the evidence for this in apical dendrites of both hippocampal cells and neocortical cells. A second mechanism is the presence of *active spots* or patches, which serve as booster zones. We have noted that these appear to be present in chromatolytic motoneurons (Chapter 5), Purkinje cells (Chapter 8), and hippocampal pyramidal cells (Chapter 11). Fast prepotentials and dendritic spikes generated by such zones have not been reported in cortical neurons of adult animals; they have been observed, however, in recording from immature cortex of very young animals (see Purpura, 1972). A third mechanism is the *potentiation* of synaptic potentials to give larger amplitude responses upon repetitive stimulation; as we have seen, this is a characteristic that synapses in the neocortex share with synapses in the hippocampus (Chapter 11); it is also found, to a certain extent, in the synapses that neocortical cells make onto spinal motoneurons (Chapter 5). Potentiation, in effect, brings a given synapse nearer, functionally, to the axon hillock, and is therefore an important means by which an input could dominate a pyramidal neuron impulse output. Whether such a mechanism is operative during normal functioning awaits further investigation.

DENDRITIC SPINES   The presence of numerous spines is commonly regarded as one of the outstanding characteristics of pyramidal cell dendrites. The pyramidal cells are by no means unique in this respect; we have seen that spines are also outstanding characteristics of olfactory granule cells (Chapter 6), cerebellar Purkinje cells (Chapter 8), thalamic relay neurons (Chapter 9), and the pyramidal cells of olfactory cortex (Chapter 10) and the hippocampus (Chapter 11). The reader may wish to review the discussions of the properties of spines in these other neurons, for possible relevance to the case of cortical spines (see also Diamond, Gray, and Yasargil, 1970).

As in other parts of the brain, cortical spines do not appear to be a means simply for increasing the synaptic surface of a dendrite; where spines are present, synapses are located preferentially on them and not on the intervening parts of the branch.

Cortical spines have traditionally been regarded as entirely postsynaptic in position and, hence, receptive in function. The recent evidence for dendrodendritic synapses (see Fig. 62), however, suggests that some cortical spines may occupy presynaptic positions and have output functions as well.

If the spine provides a specific synaptic site, what is the specific function of that site? Much attention has recently been focused on the possibility that the cortical spine provides a synaptic site that is especially subject to modification by use, i.e., by sensory input or learning experience. In kittens, deprivation of visual input in the early weeks of life has been reported to lead to deformation (Globus and Schiebel, 1967) and loss (Valverde, 1967) of spines. The spines affected are on the proximal shafts of apical dendrites in layer IV, i.e., in the layer in which the specific afferent terminals from the lateral geniculate nuclei are located. It has been suggested that these changes reflect the sensitivity of the immature developing cortex to changes in the external environment. Physiological experiments have shown that the functional properties of cortical neurons depend on early sensory input (e.g., Blakemore and Cooper, 1970). Further evidence is needed to correlate the spine changes with the functional effects.

If spines provide a modifiable synaptic substrate, it must be asked what kind of modification is involved? This brings us back to an assessment of electrotonic properties. Rall and Rinzel (1971) have pointed out the spine stem is the critical site for the control of the electrotonic relation between the spine head (the synaptic site) and the branch to which it is connected. The common variations in the morphology of this relation are illustrated in Fig. 67. In (A) is a very stubby spine, which is characteristic of spines found on the large diameter apical shaft and dendritic branches. An intermediate case is shown in (B). In (C), a spine with a long thin stem arises from a distal dendritic branch of small caliber.

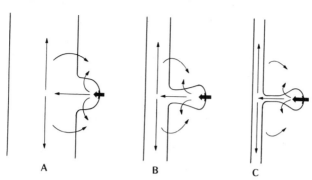

FIG. 67. Diagrams to illustrate electrotonic current flows generated by synaptic inputs to different types of dendritic spines. (A) Stubby spine arising from a thick branch; (B) Moderately elongated spine from a medium-sized branch; (C) Elongated spine with a thin stem arising from a thin branch. (After Rall and Rinzel, 1971.)

The amount of spread of a synaptic potential from the spine to the branch is governed by the ratio of the spine stem resistance to the input resistance of the branch. Rall and Rinzel (1971) have shown that, when this ratio is very large or very small, changes in spine stem caliber (and, hence, resistance) have little effect on synaptic potential spread to the branch. In the middle range, however, when the ratio is near one, a small change in the spine stem resistance has a relatively large effect on this spread. "Over this favorable range . . . fine adjustments of the stem resistances of many spines, as well as changes in dendritic caliber . . . could provide an organism with a way to adjust the relative weights of the many synaptic inputs received by such neurons; this could contribute to plasticity and learning of a nervous system" (Rall and Rinzel, 1971). Preliminary study suggests that the cortical spines and branches have dimensions that are, in fact, what would be expected to provide optimal conditions for these mechanisms to occur (see Fig. 67; see also Rall and Rinzel, 1973).

It is of interest to note that the above considerations of electrotonic current spread are also relevant to the diffusion of substances through a solute; the equations for the two cases are exactly equivalent. Adjustments in spine stem caliber may, therefore, be postulated to have a critical effect on the movement of substances into and out of the spine, through which long-term control over metabolic processes in the spine could be exerted. From this point of view, the spine appears not only as a relatively independent input-output locus, but also as a device for creating a microenvironment, whose internal composition is subject to maximal effect by a synaptic action and to control through adjustments of the stem that links it to the metabolic sources of its parent neuron.

These considerations apply generally to the spines of dendrites in the nervous system; their relevance for such particular cases as the cerebellum (Chapter 8) or the cortex await further experimental evidence. It will be of particular interest to determine whether the lability of potency of synaptic actions in cortical neurons can be correlated with modifiability of the spines. Although the spines are undoubtedly crucial sites in cortical function, it seems likely that the unique contribution of the neocortex to brain function must be viewed within the context of the many other aspects of synaptic organization discussed in this chapter.

REFERENCES

Adrian, E. D. 1950. The electrical activity of the mammalian olfactory bulb. Electroencephalog. Clin. Neurophysiol. 2: 377-388.

Adrian, E. D. 1953. Sensory messages and sensation: The response of the olfactory organ to different smells. Acta physiol. scand. 29: 5-14.

Agduhr, E. 1934. Vergleich der Neuritenanzahl in dem Wurzeln der Spinalnerven bei Kröte, Maus, Hund und Mensch. Z. Anat. 102: 194-210.

Aitken, J. and J. Bridger. 1961. Neuron size and neuron population density in the lumbosacral region of cat spinal cord. J. Anat. (London) 95: 38-53.

Aitken, L. M. and C. W. Dunlop. 1969. Inhibition in the medial geniculate body of the cat. Exp. Brain Res. 7: 68-83.

Allen, R. A. 1969. The retinal bipolar cells and their synapses in the inner plexiform layer. In The Retina: Morphology, Function and Clinical Characteristics (B. R. Straatsma, M. O. Hall, R. A. Allen and F. Crescitelli, eds.). UCLA Forum in Medical Sciences No. 8. pp. 101-143.

Allison, A. C. 1953. The morphology of the olfactory system in the vertebrates. Biol. Rev. 28: 195-244.

Allison, A. C. and R. T. T. Warwick. 1949. Quantitative observations on the olfactory system of the rabbit. Brain 72: 186-197.

Amassian, V. E. and H. Weiner. 1966. Monosynaptic and polysynaptic activation of pyramidal tract neurons by thalamic stimulation. In The Thalamus (D. P. Purpura and M. D. Yahr, eds.). New York: Columbia University Press. pp. 255-282.

Andersen, P. and S. A. Andersson. 1968. Physiological Basis of the Alpha Rhythm. New York: Appleton-Century-Crofts.

Andersen, P., T. W. Blackstad, and T. Lømo. 1966. Location and identification of excitatory synapses on hippocampal pyramidal cells. Exp. Brain Res. 1: 236-248.

Andersen, P., B. H. Bland, and J. D. Dudar. 1973. Organization of the hippocampal output. Exp. Brain Res. 17: 152-168.

Andersen, P., T. V. P. Bliss, and K. K. Skrede. 1971. Lamellar organization of hippocampal excitatory pathways. Exp. Brain Res. 13: 222-238.

Andersen, P. and J. C. Eccles. 1962. Inhibitory phasing of neuronal discharge. Nature 196: 645-647.

Andersen, P., J. C. Eccles, and Y. Løyning. 1964. Location of postsynaptic synapses on hippocampal pyramids. J. Neurophysiol. 27: 592-607.

Andersen, P., J. C. Eccles, and T. A. Sears. 1964. The ventro-basal complex of the thalamus: types of cells, their responses and their functional organization. J. Physiol. (London) 174: 370-399.

Andersen, P., B. Holmquist, and P. E. Voorhoeve. 1966. Excitatory synapses on hippocampal apical dendrites activated by entorhinal stimulation. Acta physiol. scand. 66: 461-472.

Andres, K. H. 1965. Der Feinbau des Bulbus Olfactorius der Ratte unter besonderer Berücksichtigung der synaptischen Verbindungen. Z. Zellforsch. 65: 530-561.

Andres, K. H. 1970. Anatomy and ultrastructure of the olfactory bulb in fish, amphibia, reptiles, birds and mammals. *In* CIBA Foundation Symposium on Taste and Smell in Vertebrates (G. E. W. Wolstenholme and Julie Knight, eds.). pp. 177-194.

Angaut, P. and C. Sotelo. 1973. The fine structure of the cerebellar central nuclei in the cat. II. Synaptic organization. Exp. Brain Res. 16: 431-454.

Asanuma, H. and I. Rosén. 1973. Spread of mono- and polysynaptic connections within cat's motor cortex. Exp. Brain Res. 16: 507-520.

Ayala, G. F., M. Dichter, R. J. Gumnit, H. Matsumoto, and W. A. Spencer. 1973. Genesis of epileptic interictal spikes. New knowledge of cortical feedback systems suggests a neurophysiological explanation of brief paroxysms. Brain Res. 52: 1-17.

Baker, R. and R. Llinás. 1971. Electrotonic coupling between neurones in the rat mesencephalic nucleus. J. Physiol. (London) 212: 45-63.

Barlow, H. B. 1953. Summation and inhibition in the frog's retina. J. Physiol. (London). 119: 69-88.

Barret, J. N. and W. E. Crill. 1974. Specific membrane properties of cat motoneurones. J. Physiol. (London) (in press).

Baylor, D. A., M. G. F. Fuortes, and P. M. O'Bryan. 1971. Receptive fields of cones in the retina of the turtle. J. Physiol. (London) 214: 265-294.

References 331

Baylor, D. A. and J. G. Nicholls. 1969. After-effects of nerve impulses on signalling in the central nervous system of the leech. J. Physiol. (London) 203: 571-589.

Bennett, M. V. L. 1973. Function of electrotonic junctions in embryonic and adult tissues. Fed. Proc. 32: 65-75.

Berry, R. W. and M. J. Cohen. 1972. Synaptic stimulation of RNA metabolism in the giant neuron of *Aplysia californica*. J. Neurobiol. 3: 209-222.

Biedenbach, M. A. and C. F. Stevens. 1969. Synaptic organization of cat olfactory cortex as revealed by intracellular recording. J. Neurophysiol. 32: 204-214.

Bishop, G. H. 1956. Natural history of the nerve impulse. Physiol. Rev. 36: 376-399.

Blackstad, T. W. 1967. Cortical grey matter. A correlation of light and electron microscopic data. *In* The Neuron (H. Hydén, ed.). Amsterdam: Elsevier. pp. 49-118.

Blackstad, T. W. and A. Kjaerheim. 1961. Special axo-dendritic synapses in the hippocampal cortex: Electron and light microscopic studies on the layer of mossy fibers. J. Comp. Neur. 117: 133-159.

Blakemore, C. and G. F. Cooper. 1970. Development of the brain depends on the visual environment. Nature 228: 477-478.

Bliss, T. V. P. and A. R. Gardner-Medwin. 1973. Long-lasting potentiation of synaptic transmission in the dentate area of the unanaesthetized rabbit following stimulation of the perforant pathway. J. Physiol. (London) 232: 357-374.

Bliss, T. V. P. and T. Lømo. 1973. Long-lasting potentiation of synaptic transmission in the dentate area of the anaesthetized rabbit following stimulation of the perforant path. J. Physiol. (London) 232: 331-356.

Bloom, F. E., B. J. Hoffer, and G. R. Siggins. 1971. Studies on norepinephrine-containing afferents to Purkinje cells of rat cerebellum. I. Localization of the fibers and their synapses. Brain Res. 25: 501-521.

Bodian, D. 1966. Synaptic types on spinal motoneurons: An electron microscopic study. Bull. Johns Hopkins Hosp. 119: 16-45.

Bodian, D. 1972. Synaptic diversity and characterization by electron microscopy. *In* Structure and Function of Synapses (G. D. Pappas and D. P. Purpura, eds.). New York: Raven. pp. 45-66.

Boycott, B. B. and J. E. Dowling. 1969. Organization of the primate retina: Light microscopy. Phil. Trans. Roy. Soc. Lond. B. 255: 109-184.

Boycott, B. B. and H. Kolb. 1973. The horizontal cells of the rhesus monkey retina. J. Comp. Neur. 148: 115-140.

Braitenberg, V. and R. P. Atwood. 1958. Morphological observations on the cerebellar cortex. J. Comp. Neur. 109: 1-27.

Brightman, M. W. and T. S. Reese. 1969. Junctions between intimately apposed cell membranes in the vertebrate brain. J. Cell Biol. 40: 648-677.

Bruesch, S. R. and L. B. Arey. 1942. The number of myelinated and unmyelinated fibers in the optic nerve of vertebrates. J. Comp. Neur. 77: 631-665.

Brodal, A. 1947. The hippocampus and the sense of smell. A review. Brain 70: 179-222.

Brodal, A. 1969. Neurological Anatomy. New York: Oxford University Press.

Burke, R. E. 1971. Control systems operating on spinal reflex mechanisms. In Central Control of Movement (E. V. Evarts, ed.). Neurosci. Res. Prog. Bull. 9 (No. 1). pp. 60-85.

Burke, R. E. and G. ten Bruggencate. 1971. Electrotonic characteristics of alpha motoneurones of varying size. J. Physiol. (London) 212: 1-20.

Burke, W. and A. J. Sefton. 1966. Discharge patterns of principal cells in lateral geniculate nucleus of rat. J. Physiol. (London) 187: 201-212.

Cajal, S. Ramón y. 1891. Sur la Structure de l'écorce cerebrale de quelques mammifères. La Cellule 7: 124-176.

Cajal, S. Ramón y. 1911. Histologie du Système nerveux de l'homme et des vertébrés. Paris: Maloine.

Cajal, S. Ramón y. 1955. Studies on the Cerebral Cortex. London: Lloyd-Luke.

Calleja, C. 1896. La region olfactoria del cerebro. Madrid. N. Moya

Chandler, W. K. and H. Meves. 1965. Voltage clamp experiments on internally perfused giant axons. J. Physiol. (London) 180: 788-820.

Chan-Palay, V. 1971. The recurrent collaterals of Purkinje cell axons: a correlated study of the rat's cerebellar cortex with electron microscopy and the Golgi method. Z. Anat. Entwickl.-Gesch. 134: 200-234.

Chan-Palay, V. 1973. Neuronal circuitry in the nucleus lateralis of the cerebellum. Zeit. Anat. Entwickl.-Gesch. 142: 259-265.

Chan-Palay, V. and S. L. Palay. 1971a. Tendril and glomerular collaterals of climbing fibers in the granular layer of the rat's cerebellar cortex. Z. Anat. Entwickl.-Gesch. 133: 247-273.

Chan-Palay, V. and S. L. Palay. 1971b. The synapse *en marron* between Golgi II neurons and mossy fibers in the rat's cerebellar cortex. Z. Anat. Entwickl.-Gesch. 133: 274-287.

Charlton, B. T. and E. G. Gray. 1966. Comparative electron microscopy of synapses in the vertebrate spinal cord. J. Cell Sci. 1: 67-80.

Clark, W. E. le Gros. 1951. The projection of the olfactory epithelium on the olfactory bulb in the rabbit. J. Neurol. Neurosurg. Psychiat. 14: 1-10.

Clark, W. E. le Gros. 1957. Inquiries into the anatomical basis of olfactory discrimination. Proc. Roy. Soc. B 146: 299-319.

Cohen, A. I. 1972. Rods and cones. *In* Physiology of Photoreceptor Organs (M. G. F. Fuortes, ed.). Handbook of Sensory Physiology (Vol. VII/1B). Berlin: Springer. pp. 63-110.

Colonnier, M. 1968. Synaptic patterns on different cell types in the different laminae of the cat visual cortex. An electron microscopic study. Brain Res. 9: 268-287.

Colquhoun, D., R. Henderson, and J. M. Ritchie. 1972. The binding of labelled tetrodotoxin to non-myelinated nerve fibers. J. Physiol. (London) 227: 95-126.

Conradi, S. 1969. On motoneuron synaptology in adult cats. Acta physiol. scand. Suppl. 332. 1-115.

Cooper, J. R., F. E. Bloom, and R. H. Roth. 1974. The Biochemical Basis of Neuropharmacology. New York: Oxford University Press.

Cowan, W. M. 1970. Centrifugal fibres to the avian retina. Brit. Med. Bull. 26: 112-118.

Crill, W. E. 1970. Unitary multi-spiked responses in cat inferior olive nucleus. J. Neurophysiol. 33: 199-209.

Daitz, H. M. and T. P. S. Powell. 1954. Studies of the connexions of the fornix system. J. Neurol. Neurosurg. Psychiat. 17: 75-82.

Dale, H. H. 1935. Pharmacology and nerve endings. Proc. Roy. Soc. Med. 28: 319-332.

Deiters, O. 1865. Untersuchungen uber Gehirn und Ruckenmark. Braunschweig: Vieweg.

De Robertis, E. D. P., W. W. Nowinski, and F. A. Sears. 1970. Cell Biology. Philadelphia: Saunders.

Diamond, J., E. G. Gray, and G. M. Yasargil. 1970. The function of the dendritic spine: An hypothesis. *In* Excitatory Synaptic Mechanisms (P. Andersen and J. K. S. Jansen, eds.). Oslo: Universitetsforlag. pp. 213-222.

Dowling, J. E. 1968. Synaptic organization of the frog retina: An electron microscopic analysis comparing the retinas of frogs and primates. Proc. Roy. Soc. B 170: 205-228.

Dowling, J. E. 1970. Organization of vertebrate retinas. Invest. Ophth. 9: 655-680.

Dowling, J. E. and B. B. Boycott. 1966. Organization of the primate retina: Electron microscopy. Proc. Roy. Soc. B. 166: 80-111.

Dowling, J. E. and H. Ripps. 1973. Effect of magnesium on horizontal cell activity in the Skate retina. Nature 242: 101-103.

Eccles, J. C. 1953. The Neurophysiological Basis of Mind. Oxford: Clarendon Press.

Eccles, J. C. 1957. The Physiology of Nerve Cells. Baltimore: Johns Hopkins University Press.

Eccles, J. C. 1964. The Physiology of Synapses. Berlin: Springer.

Eccles, J. C. 1969. The Inhibitory Pathways of the Central Nervous System. Springfield: Thomas.

Eccles, J. C. 1973. The cerebellum as a computer: Patterns in space and time. J. Physiol. (London) 229: 1-32.

Eccles, J. C., P. Fatt, and K. Koketsu. 1955. Cholinergic and inhibitory synapses in a pathway from motor-axon collaterals to motoneurones. J. Physiol. (London) 216: 524-562.

Eccles, J. C., M. Ito, and J. Szentágothai. 1967. The Cerebellum as a Neuronal Machine. Berlin: Springer.

Eccles, J. C., B. Libet, and R. R. Young. 1958. The behavior of chromatolyzed motoneurons studied by intracellular recording. J. Physiol. (London) 143: 11-40.

Edwards, C. and D. Ottoson. 1958. The site of impulse initiation in a nerve cell of a crustacean stretch receptor. J. Physiol. (London) 143: 138-148.

Erulkar, S. D., C. W. Nichols, M. B. Popp, and G. B. Koelle. 1968. Renshaw elements: Localization and acetylcholinesterase content. J. Histochem. Cytochem. 16: 128-135.

Evarts, E. V. 1968. Relation of pyramidal tract activity to force exerted during voluntary movement. J. Neurophysiol. 31: 14-27.

Evarts, E. V. 1971. Feedback and corollary discharge: A merging of the concepts. *In* Central Control of Movement (E. V. Evarts, ed.). Neurosci. Res. Prog. Bull. 9 (No. 1). pp. 86-112.

Fadiga, E. and J. M. Brookhart. 1960. Monosynaptic activation of different portions of the motor neuron membrane. Am. J. Physiol. 198: 693-703.

Famiglietti, E. V., Jr. 1970. Dendro-dendritic synapses in the lateral geniculate nucleus of the cat. Brain Res. 20: 181-191.

Famiglietti, E. V., Jr., and A. Peters. 1972. The synaptic glomerulus and the intrinsic neuron in the dorsal lateral geniculate nucleus of the cat. J. Comp. Neur. 144: 285-334.

Fox, C. A. and J. W. Barnard. 1957. A quantitative study of the Purkinje cell dendritic branchlets and their relationship to afferent fibers. J. Anat. 91: 299-313.

Frank, K. and M. G. F. Fuortes. 1956. Stimulation of spinal motoneurones with intracellular electrodes. J. Physiol. (London) 134: 451-470.

Forbes, A. 1922. The interpretation of spinal reflexes in terms of present knowledge of nerve conduction. Physiol. Rev. 2: 361-414.

Freeman, W. J. 1964. A linear distributed feedback model for prepyriform cortex. Exp. Neurol. 10: 525-547.

Freeman, W. J. 1968. Patterns of variation in waveform of averaged evoked potentials from prepyriform cortex of cats. J. Neurophysiol. 31: 1-13.

Fujita, Y. 1968. Activity of dendrites of single Purkinje cells and its relationship to so-called inactivation response to rabbit cerebellum. J. Neurophysiol. 31: 131-141.

Garey, L. J. 1971. A light and electron microscopic study of the visual cortex of the cat and monkey. Proc. Roy. Soc. B. 179: 21-40.

Gelfan, S. 1963. Neurone and synapse populations in spinal cord: indication of role in total integration. Nature 198: 162-163.

Globus, A. and A. B. Schiebel. 1967. The effect of visual deprivation on cortical neurons: A Golgi study. Exp. Neurol. 19: 331-345.

Gottlieb, D. I. and W. M. Cowan. 1972. On the distribution of axonal terminals containing spheroidal and flattened synaptic vesicles in the hippocampus and dentate gyrus of the rat and cat. Z. Zellforsch. 129: 413-429.

Granit, R. 1955. Receptors and Sensory Perception. New Haven: Yale University Press.

Granit, R. and C. G. Phillips. 1956. Excitatory and inhibitory processes acting upon individual Purkinje cells of the cerebellum in cats. J. Physiol. (London) 133: 520-547.

Gray, E. G. 1959. Axo-somatic and axo-dendritic synapses of the cerebral cortex: An electron-microscope study. J. Anat. 93: 420-433.

Gray, E. G. 1962. A morphological basis for presynaptic inhibition? Nature 193: 82-83.

Green, J. D. 1964. The hippocampus. Physiol. Rev. 44: 561-608.

Greengaard, P., J. W. Kebabian, and D. A. McAfee. 1972. Studies on the role of cyclic AMP in neural function. Proc. 5th Int. Congr. Pharmacol. vol. 5, pp. 207-217.

Grossman, A., A. R. Lieberman, and K. E. Webster. 1973. A Golgi study of the rat dorsal lateral geniculate nucleus. J. Comp. Neur. 150: 441-466.

Guillery, R. W. 1971. Patterns of synaptic interconnections in the dorsal lateral geniculate nucleus of cat and monkey: A brief review. Vision Res. Suppl. no. 3: 211-227.

Haberly, L. B. 1969. Single unit responses to odor in the prepyriform cortex of the rat. Brain Res. 12: 481-484.

Haberly, L. B. 1973a. Unitary analysis of opossum prepyriform cortex. J. Neurophysiol. 36: 762-787.

Haberly, L. B. 1973b. Summed potentials evoked in opossum prepyriform cortex. J. Neurophysiol. 36: 775-788.

Haberly, L. B. and G. M. Shepherd. 1973. Current density analysis of summed evoked potentials in opossum prepyriform cortex. J. Neurophysiol. 36: 789-802.

Hagins, W. A., R. D. Penn, and S. Yoshikami. 1970. Dark current and photocurrent in retinal rods. Biophys. J. 10: 380-412.

Hamlyn, L. H. 1963. An electron microscope study of pyramidal neurons in the Ammon's horn of the rabbit. J. Anat. 97: 189-201.

Harding, B. N. 1971. Dendro-dendritic synapses, including reciprocal synapses, in the ventrolateral nucleus of the monkey thalamus. Brain Res. 34: 181-185.

Hebb, D. O. 1961. The Organization of Behavior. New York: Wiley.

Hecht, S., S. Shlaer and M. H. Pirenne. 1942. Energy, quanta and vision. J. Gen. Physiol. 25: 819-840.

Heimer, L. 1968. Synaptic distribution of centripetal and centrifugal nerve fibres in the olfactory system of the rat. An experimental anatomical study. J. Anat. 103: 413-432.

Henneman, E. 1968. Organization of the spinal cord. *In* Medical Physiology (V. B. Mountcastle, ed.) vol. II. St. Louis: Saunders. pp. 1717-1732.

Herrick, C. J. 1948. The Brain of the Tiger Salamander. Chicago: Chicago University Press.

Heuser, J. E. and T. S. Reese. 1973. Evidence for recycling of synaptic vesicle membrane during transmitter release at the frog neuromusclar junction. J. Cell Biol. 57: 315-344.

Hild, W. and I. Tasaki. 1962. Morphological and physiological properties of neurons and glial cells in tissue culture. J. Neurophysiol. 25: 277-304.

Hirata, Y. 1964. Some observations on the fine structure of the synapses in the olfactory bulb of the mouse, with particular reference to the ayptical synaptic configuration. Arch. Histol. Japan. 24: 293-302.

Hjorth-Simonsen, A. 1973. Some intrinsic connections of the hippocampus in the rat: an experimental analysis. J. Comp. Neur. 147: 145-162.

Hjorth-Simonsen, A. and B. Jeune. 1972. Origin and termination of the hippocampal perforant path in the rat studied by silver impregnation. J. Comp. Neur. 144: 215-232.

Hodgkin, A. L. 1964. The Conduction of the Nervous Impulse. Springfield: Thomas.

Hodgkin, A. L. 1972. Recent work on visual mechanisms. Proc. Roy. Soc. B. 180: X-XX.

Hodgkin, A. L. and A. F. Huxley. 1952. A quantitative description of membrane current and its application to conduction and excitation in nerve. J. Physiol. (London) 117: 500-544.

Hubbard, J. I., R. Llinás, and D. M. J. Quastel. 1969. Electrophysiological Analysis of Synaptic Transmission. Baltimore: Williams & Wilkins.

Hubel, D. H. and Wiesel. 1962. Receptive fields, binocular interaction and functional architecture in the cat's visual cortex. J. Physiol. (London) 160: 106-154.

Hubel, D. H. and T. Wiesel. 1968. Receptive fields and functional architecture of monkey striate cortex. J. Physiol. (London) 195: 215-243.

Hultborn, H., E. Jankowska, and S. Lindström. 1971. Recurrent inhibition of interneurones monosynaptically activated from group Ia afferents. J. Physiol. (London) 215: 613-636.

Iversen, L. L. 1970. Neurotransmitters, neurohormones and other small molecules in neurons. In The Neurosciences: Second Study Program (F. O. Schmitt, ed.-in-chief) New York: Rockefeller. pp. 768-781.

Jack, J. J. B., S. Miller, R. Porter, and S. J. Redman. 1971. The time course of minimal excitatory post-synaptic potentials evoked in spinal motoneurones by group Ia afferent fibers. J. Physiol. (London) 215: 353-380.

Jack, J. J. B. and S. J. Redman. 1971. An electrical description of the motoneurone, and its application to the analysis of synaptic potentials. J. Physiol. (London) 215: 321-352.

Jacobson, S. and D. A. Pollen. 1968. Electrotonic spread of dendritic potentials in feline pyramidal cells. Science 164: 1351-1353.

Jankowska, E. and S. Lindström. 1971. Morphological identification of Renshaw cells. Acta physiol. scand. 81: 428-430.

Jankowska, E. and S. Lindström. 1972. Morphology of interneurones mediating Ia reciprocal inhibition of motoneurones in the spinal cord of the cat. J. Physiol. (London) 226: 805-823.

Jansen, J. K. S. and J. G. Nicholls. 1973. Conductance changes, an electrogenic pump and the hyperpolarization of leech neurones following impulses. J. Physiol. (London) 229: 635-655.

Jansen, J. K. S. and L. Walløe. 1970. Signal transmission between successive neurons in the dorsal spinocerebellar pathway. In The Neurosciences: Second Study Program (F. O. Schmitt, ed.-in-chief). New York: Rockefeller. pp. 617-629.

Jasper, H. H., A. A. Ward, and A. Pope. 1969. Basic Mechanisms of the Epilepsies. Boston: Little, Brown.

Jones, E. G. and T. P. S. Powell. 1969. Electron microscopy of synaptic glomeruli in the thalamic relay nuclei of the cat. Proc. Roy. Soc. B 172: 153-171.

Jones, E. G. and T. P. S. Powell. 1970a. Electron microscopy of the somatic sensory cortex of the cat. I. Cell types and synaptic organization. Phil. Trans. Roy. Soc. B 257: 1-11.

Jones, E. G. and T. P. S. Powell. 1970b. An electron microscopic study of the laminar pattern and mode of termination of afferent fibre pathways in the somatic sensory cortex of the cat. Phil. Trans. Roy. Soc. B 257: 45-62.

Kalil, R. E. and R. Chase. 1970. Corticofugal influence on activity of lateral geniculate neurons in the cat. J. Neurophysiol. 33: 459-474.

Kandel, E. R. and W. A. Spencer. 1961. Electrophysiology of hippocampal neurons. II. After-potentials and repetitive firing. J. Neurophysiol. 24: 243-259.

Kandel, E. R., W. A. Spencer, and F. J. Brinley, Jr. 1961. Electrophysiology of hippocampal neurons. I. Sequential invasion and synaptic organization. J. Neurophysiol. 24: 225-242.

Kandel, E. R. and W. A. Spencer. 1968. Cellular neurophysiological approaches in the study of learning. Physiol. Rev. 48: 65-134.

Kane, E. C. 1973. Octopus cells in the cochlear nucleus of the cat: heterotypic synapses upon homeotypic neurons. Intern. J. Neuroscience 5: 251-279.

Kaneko, A. 1970. Physiological and morphological identification of horizontal, bipolar and amacrine cells in goldfish retina. J. Physiol. (London) 207: 623-633.

Katz, B. 1966. Nerve, Muscle and Synapse. New York: McGraw-Hill.

Katz, B. and R. Miledi. 1967. A study of synaptic transmission in the absence of nerve impulses. J. Physiol. (London) 192: 407-436.

Kernell, D. and R. P. Petersen. 1970. The effect of spike activity versus synaptic activation on the metabolism of ribonucleic acid in a molluscan giant neuron. J. Neurochem. 17: 1087-1094.

Kidd, M. 1962. Electron microscopy of the inner plexiform layer of the retina in the cat and pigeon. J. Anat. 96: 179-187.

Kolb, H. 1970. Organization of the outer plexiform layer of the primate retina: electron microscopy of Golgi-impregnated cells. Proc. Roy. Soc. B. 258: 261-283.

Krnjevic, K. 1970. Central excitatory transmitters in vertebrates. In Excitatory Synaptic Mechanisms (P. Andersen and J. K. S. Jansen, eds.). Oslo: Universitetsforlag. pp. 95-104.

Kuffler, S. W. 1953. Discharge patterns and functional organization of mammalian retina. J. Neurophysiol. 16: 37-68.

Kuno, M. 1971. Quantum aspects of central and ganglionic synaptic transmission in vertebrates. Physiol. Rev. 51: 647-678.

Kuno, M. and R. Llinás. 1970. Alterations of synaptic action in chromatolysed motoneurones of the cat. J. Physiol. (London) 210: 823-838.

Laatsch, R. H. and W. M. Cowan. 1966. Electron microscopic studies of the dentate gyrus of the rat. I. Normal structure with special reference to synaptic organization. J. Comp. Neur. 128: 359-396.

Ladman, A. J. 1958. The fine structure of the rod-bipolar cell synapse in the retina of the albino rat. J. Biophys. Biochem. Cytol. 4: 459-466.

Landgren, S., C. G. Phillips, and R. Porter. 1962. Cortical fields of origin of the monosynaptic pyramidal pathways to some alpha motoneurones of the baboon's hand and forearm. J. Physiol. (London) 161: 112-125.

Landis, D. M. D. and T. S. Reese. 1974. Differences in membrane structure between excitatory and inhibitory synapses in the cerebellar cortex. J. Comp. Neur. (in press).

LeVay, S. 1971. On the neurons and synapses of the lateral geniculate nucleus of the monkey, and the effects of eye enucleation. Z. Zellforsch. 113: 396-419.

LeVay, S. 1973. Synaptic patterns in the visual cortex of the cat and monkey. Electron microscopy of Golgi preparations. J. Comp. Neur. 150: 53-86.

Leveteau, J. and P. MacLeod. 1966. Olfactory discrimination in the rabbit olfactory glomerulus. Science 153: 175-176.

Lieberman, A. R. and K. E. Webster. 1973. Presynaptic dendrites, dendritic boutons and reciprocal synapses in the dorsal lateral geniculate nucleus of the rat. J. Neurocytol. 2: (in press).

Llinás, R., R. Baker, and C. Sotelo 1973. Electrical transmission between cells in the inferior olive of the cat. Soc. Neurosci. Abstracts. 3: 156.

Llinás, R. and D. E. Hillman. 1969. Physiological and morphological organization of the cerebellar circuits in various vertebrates. *In* Neurobiology of Cerebellar Evolution and Development (R. Llinás, ed.). Chicago: Am. Med. Assoc. pp. 43-73.

Llinás, R. and C. Nicholson. 1971. Electrophysiological properties of dendrites and somata in alligator Purkinje cells. J. Neurophysiol. 34: 532-551.

Lloyd, D. P. C. 1943. Reflex action in relation to patterns and peripheral source of afferent stimulation. J. Neurophysiol. 6: 111-120.

Lorente de Nó, R. 1922. La corteza cerebral del raton. Trab. Lab. Invest. Biol. (Madrid) 20: 41-78.

Lorente de Nó, R. 1934. Studies on the structure of the cerebral cortex. II. Continuation of the study of the Ammonic system. J. Psychol. Neurol. 46: 113-177.

Lorente de Nó, R. 1938. The cerebral cortex: Architecture, intracortical connections and motor projections. *In* Physiology of the Nervous System (J. F. Fulton). London: Oxford University Press. pp. 291-325.

Lorente de Nó, R. and C. A. Condouris. 1959. Decremental conduction in peripheral nerve. Integration of stimuli in the neuron. Proc. Nat. Acad. Sci. 45: 592-617.

Lund, J. S. 1973. Organization of neurons in the visual cortex, area 17, of the monkey (*Macaca mulatta*). J. Comp. Neur. 147: 455-496.

Lund, R. D. and L. E. Westrum. 1966. Synaptic vesicle differences after primary formalin fixation. J. Physiol. (London) 185: 7-9P.

Lundberg, A. 1970. The excitatory control of the Ia inhibitory pathway. *In* Excitatory Synaptic Mechanisms (P. Andersen and J. K. S. Jansen, eds.). Oslo: Universitetsforlag. pp. 333-340.

Lux, H. D. and D. A. Pollen. 1966. Electrical constants of neurons in the motor cortex of the cat. J. Neurophysiol. 29: 207-220.

Lux, H. D., P. Schubert and G. W. Kreutzberg. 1970. Direct matching of morphological and electrophysiological data in cat spinal motoneurones. *In* Excitatory Synaptic Mechanisms (P. Andersen and J. K. S. Jansen, eds.). Oslo: Universitetsforlag. pp. 189-198.

MacLean, P. D. 1972. Implications of microelectrode findings on exteroceptive inputs to the limbic cortex. *In* Limbic System Mechanisms and Autonomic Function (C. H. Hockman, ed.). Springfield: Thomas. pp. 115-136.

Marmarelis, P. Z. and K. Naka. 1972. Spatial distribution of potential in a flat cell. Application to the catfish horizontal cell layers. Biophys. J. 12: 1515-1532.

Marin-Padilla, M. 1972. Double origin of the pericellular baskets of the pyramidal cells of the human motor cortex: a Golgi study. Brain Res. 38: 1-12.

Marr, D. 1969. A theory of cerebellar cortex. J. Physiol. (London) 202: 437-470.

Martin, A. R. 1966. Quantal nature of synaptic transmission. Physiol. Rev. 46: 51-66.

Matsumoto, H. and C. Ajmone-Marsan. 1964. Cortical cellular phenomena in experimental epilepsy: interictal manifestations. Exp. Neurol. 9: 286-304.

Matthews, P. B. C. 1972. Mammalian Muscle Receptors and their Central Actions. Baltimore: Williams & Wilkins.

Matthews, M. A., W. D. Willis, and V. F. Williams. 1971. Dendrite bundles in lamina IX of cat spinal cord: a possible source for electrical interaction between motoneurons? Anat. Rec. 171: 313-328.

Maturana, H. R., J. Y. Lettvin, W. S. McCulloch, and W. H. Pitts. 1960. Anatomy and physiology of vision of the frog (*Rana pipiens*). J. Gen. Physiol. 43: 129-175.

McIlwain, J. T. and O. D. Creutzfeldt. 1967. Microelectrode study of synaptic excitation and inhibition in the lateral geniculate nucleus of the cat. J. Neurophysiol. 30: 1-21.

McLardy, T. 1962. Zinc enzymes and the hippocampal mossy fibre system. Nature 194: 300-302.

McLaughlin, B. J. 1972a. The fine structure of neurons and synapses in the motor nuclei of the cat spinal cord. J. Comp. Neur. 144: 429-460.

McLaughlin, B. J. 1972b. Dorsal root projections to the motor nuclei in the cat spinal cord. J. Comp. Neur. 144: 461-474.

Mendell, L. M. and E. Henneman. 1968. Terminals of single Ia fibers: distribution within a pool of 300 homonymous motor neurons. Science 160: 96-98.

Merton, P. A. 1964. Human position sense and sense of effort. Symp. Soc. Exp. Biol. 18: 387-400.

Michael, C. R. 1969. Retinal processing of visual images. Sci. Am. 220: 104-114.

Missotten, L. 1965. The Ultrastructure of the Human Retina. Brussel: Arscia Uitgaven.

Morest, D. K. 1971. Dendrodendritic synapses of cells that have axons: the fine structure of the Golgi type II cell in the medial geniculate body of the cat. Z. Anat. Entwickl.-Gesch. 133: 216-246.

Mountcastle, V. B. 1957. Modality and topographic properties of single neurons of cat's somatic sensory cortex. J. Neurophysiol. 20: 408-434.

Mountcastle, V. B. and I. Darien-Smith. 1968. Neural mechanisms in somaesthesia. *In* Medical Physiology (V. B. Mountcastle, ed.). pp. 1372-1423.

Mountcastle, V. B. and T. P. S. Powell. 1959. Central neural mechanisms subserving position sense and kinaesthesia. Bull. Johns Hopkins Hosp. 105: 201-230.

Mugnaini, E. 1970. Neurones as synaptic targets. *In* Excitatory Synaptic Mechanisms (P. Andersen and J. K. S. Jansen, eds.). Oslo: Universitetsforlag. pp. 149-169.

Murphy, J. T., W. A. MacKay, and F. Johnson. 1973. Responses of cerebellar cortical neurons to dynamic proprioceptive interactions in the cat. Brain Res. 36: 711-723.

Nacimiento, A. C., H. D. Lux, and O. D. Creutzfeldt. 1964. Postsynaptische Potentiale von Nervenzellen des motorischen Cortex nach elektrische Reizung specifischer und unspecifischer Thalamuskerne. Arch. Ges. Physiol. 281: 152-169.

Nafstad, P. H. J. 1967. An electron microscope study on the termination of the perforant path fibers in the hippocampus and the fascia dentata. Z. Zellforsch. 76: 532-542.

Nagasawa, J., W. W. Douglas, and R. A. Schulz. 1971. Micropinocytotic origin of coated and smooth microvesicles ("synaptic vesicles") in neurosecretory terminals of posterior pituitary glands demonstrated by incorporation of horseradish peroxidase. Nature 232: 341-342.

Naka, K. and W. H. Rushton. 1966. S-potentials from colour units in the retina of fish (*Cyprinidae*). J. Physiol. (London) 185: 536-555.

Nauta, W. J. H. and H. J. Karten. 1970. A general profile of the vertebrate brain, with sidelights on the ancestry of cerebral cortex. *In* The Neurosciences: Second Study Program (F. O. Schmitt, ed.-in-chief). New York: Rockefeller. pp. 7-25.

Nelson, P. G. and S. D. Erulkar. 1963. Synaptic mechanisms of excitation and inhibition in the central auditory pathway. J. Neurophysiol. 26: 908-923.

Nelson, R. 1973. A comparison of electrical properties of neurons in *Necturus* retina. J. Neurophysiol. 36: 519-535.

Nicholls, J. G. and D. Purves. 1972. A comparison of chemical and electrical synaptic transmission between single sensory cells and a motoneurone in the central nervous system of the leech. J. Physiol. (London) 225: 637-656.

Nicoll, R. A. 1969. Inhibitory mechanisms in the rabbit olfactory bulb: Dendrodendritic mechanisms. Brain Res. 14: 157-172.

Nieuwenhuys, R. and C. Nicholson. 1969. Aspects of the histology of the cerebellum of mormyrid fishes. *In* Neurobiology of Cerebellar Evolution and Development (R. Llinás, ed.). Chicago: Am. Med. Assoc. pp. 135-169.

Noble, D., J. J. B. Jack, and R. Tsien. 1974. Electric Current Flow in Excitable Cells. Oxford: Clarendon Press.

O'Leary, J. L. 1937. Structure of the primary olfactory cortex of the mouse. J. Comp. Neur. 67: 1-31.

Olson, L. and K. Fuxe. 1971. On the projections from the locus coeruleus noradrenaline neurons: the cerebellar innervation. Brain Res. 28: 165-171.

Oscarsson, O., I. Rosén, and I. Sulg. 1966. Organization of neurones in the cat cerebral cortex that are influenced from group I muscle afferents. J. Physiol. (London) 183: 189-210.

Oshima, T. 1969. Studies of pyramidal tract cells. *In* Basic Mechanisms of the Epilepsies (H. H. Jasper, A. A. Ward, and A. Pope, eds.). Boston: Little, Brown. pp. 253-262.

Ottoson, D. and G. M. Shepherd. 1972. Transducer properties and integrative mechanisms in the frog's muscle spindle. *In* Principles of Receptor Physiology (W. R. Lowenstein, ed.). Handbook of Sensory Physiology, vol. I. New York: Springer. pp. 442-499.

Palay, S. L. 1967. Principles of cellular organization in the nervous system. *In* The Neurosciences: A Study Program (G. C. Quarton, T. Melnechuck, and F. O. Schmitt, eds.). New York: Rockefeller. pp. 24-31.

Palay, S. L. and V. Chan-Palay. 1973. Cerebellar Cortex: Cytology and Organization. Berlin: Springer.

Palkovitz, M., P. Magyar, and J. Szentágothai. 1971. Quantitative histological analysis of the cerebellar cortex in the cat. I. Number and arrangement in space of the Purkinje cells. Brain Res. 32: 1-14.

Pasik, P., T. Pasik, J. Hámori, and J. Szentágothai. 1973. Golgi type II interneurons in the neuronal circuit of the monkey lateral geniculate nucleus. Exp. Brain Res. 17: 18-34.

Penn, R. D. and W. A. Hagins. 1972. Kinetics of the photocurrent of retinal rods. Biophys. J. 12: 1073-1094.

Peters, A. and I. R. Kaiserman-Abramof. 1969. The small pyramidal neuron of rat cerebral cortex. The synapses upon dendritic spines. Z. Zellforsch. 100: 487-506.

Peters, A. and S. L. Palay. 1966. The morphology of laminae A and A_1 of the dorsal nucleus of the lateral geniculate body of the cat. J. Anat. 100: 451-486.

Peters, A., S. L. Palay and H. de F. Webster. 1970. The Fine Structure of the Nervous System. New York: Harper & Row.

Peters, A., C. C. Proskauer, and I. R. Kaiserman-Abramof. 1968. The small pyramidal neuron of the rat cerebral cortex. The axon hillock and initial segment. J. Cell Biol. 39: 604-619.

Phillips, C. G. 1956. Intracellular records from Betz cells in the cat. Quart. J. Exp. Physiol. 41: 58-69.

Phillips, C. G. 1959. Actions of antidromic pyramidal volleys on single Betz cells in the cat. Quart. J. Exp. Physiol. 44: 1-25.

Phillips, C. G. 1969. Motor apparatus of the baboon's hand. Proc. Roy. Soc. B. 173: 141-174.

Phillips, C. G. 1971. Evolution of the corticospinal tract in primates with special reference to the hand. Proc. 3rd Int. Congr. Primat. 2:2-23.

Phillips, C. G. and R. Porter. 1964. The pyramidal projections to motoneurones of some muscle groups of the baboon's forelimb. Prog. Brain Res. 12: 222-242.

Phillips, C. G., T. P. S. Powell, and G. M. Shepherd. 1963. Responses of mitral cells to stimulation of the lateral olfactory tract in the rabbit. J. Physiol. (London) 168: 65-88.

Pinching, A. J. 1971. Myelinated dendritic segments in the monkey olfactory bulb. Brain Res. 29: 133-138.

Pinching, A. J. and T. P. S. Powell. 1971. The neuropil of the glomeruli of the olfactory bulb. J. Cell Sci. 9: 347-377.

Polyak, S. L. 1941. The Retina. Chicago: Chicago University Press.

Porter, R. and J. Hore. 1969. Time course of minimal corticomotoneuronal excitatory postsynaptic potentials in lumbar motoneurones of the monkey. J. Neurophysiol. 32: 443-451.

Powell, T. P. S., R. W. Guillery, and W. M. Cowan. 1957. A quantitative study of the fornix-mammillo-thalamic system. J. Anat. 91: 419-437.

Pribram, K. H. and L. Kruger. 1954. Functions of the "Olfactory Brain." Ann. N.Y. Acad. Sci. 58: 109-138.

Price, J. L. 1973. An autoradiographic study of complementary laminar patterns of termination of afferent fibers to the olfactory cortex. J. Comp. Neur. 150: 87-108.

Price, J. L. and T. P. S. Powell. 1970a. The synaptology of the granule cells of the olfactory bulb. J. Cell Sci. 7: 125-155.

Price, J. L. and T. P. S. Powell. 1970b. An electron-microscopic study of the termination of the afferent fibres to the olfactory bulb from the cerebral hemispheres. J. Cell Sci. 7: 157-187.

Price, J. L. and T. P. S. Powell. 1971. Certain observations on the olfactory pathway. J. Anat. 110: 105-126.

Powell, T. P. S., W. M. Cowan, and G. Raisman. 1965. The central olfactory connexions. J. Anat. 99: 791-813.

Purpura, D. P. 1967. Comparative physiology of dendrites. In The Neurosciences: A Study Program (G. C. Quarton, T. Melnechuck, and F. O. Schmitt, eds.). New York: Rockefeller. pp. 372-392.

Purpura, D. P. 1972. Intracellular studies of synaptic organization in the mammalian brain. In Structure and Function of Synapses (G. D. Pappas and D. P. Purpura, eds.). New York: Raven. pp. 257-302.

Purpura, D. P. and B. Cohen. 1962. Intracellular recording from thalamic neurons during recruiting responses. J. Neurophysiol. 25: 621-635.

Purpura, D. P., R. J. Shofer, and F. S. Musgrave. 1964. Cortical intracellular potentials during augmenting and recruiting responses.

II. Patterns of synaptic activities in pyramidal and nonpyramidal tract neurons. J. Neurophysiol. 27: 133-151.

Rafols, J. A. and F. Valverde. 1973. The structure of the dorsal lateral geniculate nulceus in the mouse. A Golgi and electron microscopic study. J. Comp. Neur. 150: 303-332.

Raisman, G. 1969. Neuronal plasticity in the septal nuclei of the adult rat. Brain Res. 14: 25-48.

Raisman, G., W. M. Cowan, and T. P. S. Powell. 1965. The extrinsic afferent, commissural and association fibres of the hippocampus. Brain 88: 963-996.

Rakić, P. and R. L. Sidman. 1973. Organization of cerebellar cortex secondary to deficit of granule cells in Weaver mutant mice. J. Comp. Neur. 139: 473-500.

Rall, W. 1957. Membrane time constant of motoneurons. Science 126: 454-455.

Rall, W. 1959a. Dendritic current distribution and whole neuron properties. Naval Med. Res. Inst., Research Report NM 01-05-00.01.02.

Rall, W. 1959b. Branching dendritic trees and motoneuron membrane resistivity. Exp. Neurol. 1: 491-527.

Rall, W. 1960. Membrane potential transients and membrane time constant of motoneurons. Exp. Neurol. 2: 503-532.

Rall, W. 1962a. Theory of physiological properties of dendrites. Ann. N.Y. Acad. Sci. 96: 1071-1092.

Rall, W. 1962b. Electrophysiology of a dendritic neuron model. Biophys. J. 2 (Suppl): 145-167.

Rall, W. 1964. Theoretical significance of dendritic trees for neuronal input-output relations. *In* Neural Theory and Modelling (R. F. Reiss, ed.). Palo Alto: Stanford University Press. pp. 73-97.

Rall, W. 1967. Distinguishing theoretical synaptic potentials computed for different soma-dendritic distributions of synaptic input. J. Neurophysiol. 30: 1138-1168.

Rall, W. 1969. Time constants and electrotonic lengths of membrane cylinders and neurons. Biophys. J. 9: 1483-1508.

Rall, W. 1970a. Dendritic neuron theory and dendrodendritic synapses in a simple cortical system. *In* The Neurosciences: Second Study Program (F. O. Schmitt, ed.-in-chief). New York: Rockefeller. pp. 552-565.

Rall, W. 1970b. Cable properties of dendrites and effects of synaptic location. *In* Excitatory Synaptic Mechanisms (P. Andersen and J. K. S. Jansen, eds.). Oslo: Universitetsforlag. pp. 175-187.

Rall, W., R. E. Burke, T. G. Smith, P. G. Nelson, and K. Frank. 1967. Dendritic location of synapses and possible mechanisms for the monosynaptic EPSP in motoneurons. J. Neurophysiol. 1967: 1169-1193.

Rall, W. and J. Rinzel. 1971. Dendritic spine function and synaptic attenuation calculations. Soc. Neurosci. Abstracts. 64.

Rall, W. and J. Rinzel. 1973. Branch input resistance and steady attenuation for input to one branch of a dendritic neuron model. Biophys. J. 13: 648-688.

Rall, W. and G. M. Shepherd. 1968. Theoretical reconstruction of field potentials and dendrodendritic synaptic interactions in olfactory bulb. J. Neurophysiol. 31: 884-915.

Rall, W., G. M. Shepherd, T. S. Reese, and M. W. Brightman. 1966. Dendro-dendritic synaptic pathway for inhibition in the olfactory bulb. Exp. Neurol. 14: 44-56.

Ralston, H. J., III. 1968. The fine structure of neurons in the dorsal horn of the cat spinal cord. J. Comp. Neur. 132: 275-302.

Ralston, H. J., III. 1971. Evidence for presynaptic dendrites and a proposal for their mechanism of action. Nature 230: 585-587.

Ralston, H. J., III, and M. M. Herman. 1969. The fine structure of neurons and synapses in the ventrobasal thalamus of the cat. Brain Res. 14: 77-97.

Ramón-Moliner, E. 1962. An attempt at classifying nerve cells on the basis of their dendritic patterns. J. Comp. Neur. 119: 211-227.

Reese, T. S. and M. W. Brightman. 1970. Olfactory surface and central olfactory connections in some vertebrates. *In* CIBA Foundation Symposium on Taste and Smell in Vertebrates (G. E. W. Wolstenholme and Julie Knight, eds.). London: Churchill. pp. 115-149.

Reese, T. S. and G. M. Shepherd. 1972. Dendro-dendritic synapses in the central nervous system. *In* Structure and Function of Synapses (G. D. Pappas and D. P. Purpura, eds.). New York: Raven. pp. 121-136.

Rexed, B. 1954. A cytoarchitectonic atlas of the spinal cord in the cat. J. Comp. Neur. 100: 297-379.

Ritchie, J. M. 1971. Electrogenic ion pumping in nervous tissue. *In* Current Topics in Bioenergetics (vol. 4). pp. 327-356.

Ryall, R. W., M. F. Piercey, and C. Polosa. 1971. Intersegmental and intrasegmental distribution of mutual inhibition of Renshaw cells. J. Neurophysiol. 34: 700-707.

Schaffer, K. 1892. Beitrag zur Histologie der Ammonshornformation. Arch. Mikr. Anat. 39: 611-632.

Schiebel, M. E., T. L. Davies, and A. B. Schiebel. 1972. An unusual axonless cell in the thalamus of the adult cat. Exp. Neurol. 36: 512-518.

Schiebel, M. E. and A. B. Schiebel. 1966. Spinal motoneurons, interneurons and Renshaw cells. A Golgi study. Arch. Ital. Biol. 104: 328-353.

Schiebel, M. E. and A. B. Schiebel. 1970. Elementary processes in selected thalamic and cortical subsystems—the structural substrates. *In* The Neurosciences: Second Study Program (F. O. Schmitt, ed.-in-chief). New York: Rockefeller. pp. 443-457.

Schultz, S. G. and P. F. Curran. 1970. Coupled transport of sodium and organic solutes. Physiol. Rev. 50: 637-718.

Shepherd, G. M. 1963. Neuronal systems controlling mitral cell excitability. J. Physiol. (London) 168: 101-117.

Shepherd, G. M. 1970. The olfactory bulb as a simple cortical system: Experimental analysis and functional implications. *In* The Neurosciences: Second Study Program (F. O. Schmitt, ed.-in-chief). New York: Rockefeller. pp. 539-552.

Shepherd, G. M. 1971. Physiological evidence for dendrodendritic synaptic interactions in the rabbit's olfactory glomerulus. Brain Res. 32: 212-217.

Shepherd, G. M. 1972a. Synaptic organization of the mammalian olfactory bulb. Physiol. Rev. 52: 864-917.

Shepherd, G. M. 1972b. The neuron doctrine: a revision of functional concepts. Yale J. Biol. Med. 45: 584-599.

Sherrington, C. S. 1906. The Integrative Action of the Nervous System. New Haven: Yale University Press.

Sholl, D. 1956. The Organization of the Cerebral Cortex. London: Methuen.

Sjöstrand, F. 1958. Ultrastructure of retinal rod synapses of the guinea pig eye as revealed by three-dimensional reconstructions from serial sections. J. Ultrastruct. Res. 2: 122-170.

Sloper, J. J. 1971. Dendrodendritic synapses in the primate motor cortex. Brain Res. 34: 186-192.

Sotelo, C. and J. Taxi. 1970. Ultrastructural aspects of electrotonic junctions in the spinal cord of the frog. Brain Res. 17: 137-141.

Spencer, W. A. and E. R. Kandel. 1961a. Electrophysiology of hippocampal neurons. III. Firing level and time constant. J. Neurophysiol. 24: 260-271.

Spencer, W. A. and E. R. Kandel. 1961b. Electrophysiology of hippocampal neurons. IV. Fast prepotentials. J. Neurophysiol. 24: 272-285.

Spencer, W. A. and E. R. Kandel. 1968. Cellular and integrative properties of the hippocampal pyramidal cell and the comparative electrophysiology of cortical neurons. Int. J. Neurol. 3-4: 267-296.

Sprague, J. M. 1958. The distribution of dorsal root fibres on motor cells in the lumbosacral spinal cord of the cat, and the site of excitatory and inhibitory terminals in monosynaptic pathways. Proc. Roy. Soc. B 149: 534-556.

Stefanis, C. and H. Jasper. 1964. Intracellular microelectrode studies of antidromic responses in cortical pyramidal tract neurons. J. Neurophysiol. 27: 828-854.

Stell, W. K. 1964. Correlated light and electron microscope observations on Golgi preparations of goldfish retina. J. Cell Biol. 23: 89 A.

Stell, W. K. 1967. The structure and relationships of horizontal cells and photoreceptor-bipolar synaptic complexes in goldfish retina. Am. J. Anat. 121: 401-424.

Stell, W. K. 1972. The morphological organization of the vertebrate retina. *In* Physiology of Photoreceptor Organs (M. G. F. Fuortes, ed.). Handbook of Sensory Physiology. VII/1B. Berlin: Springer. pp. 111-214.

Stell, W. K. and P. Witkovsky. 1973. Retinal structure in the smooth dogfish, *Mustelus canis:* general description and light microscopy of giant ganglion cells. J. Comp. Neur. 148: 1-32.

Stevens, C. F. 1969. Structure of cat frontal olfactory cortex. J. Neurophysiol. 32: 184-192.

Storm-Mathisen, J. and T. W. Blackstad. 1964. Cholinesterase in the hippocampal region. Distribution and relation to architectonics and afferent systems. Acta Anat. 56: 216-253.

Svaetichin, G. 1953. The cone action potential. Acta physiol. scand. 29(Suppl 106): 565-600.

Swett, J. E. and C. M. Bourassa. 1967. Short latency activation of pyramidal tract cells by Group I afferent volleys in the cat. J. Physiol. (London) 189: 101-117.

Szentágothai, J. 1963. The structure of the synapse in the lateral geniculate body. Acta Anat. 55: 166-185.

Szentágothai, J. 1969. Architecture of the cerebral cortex. *In* Basic Mechanisms of the Epilepsies (H. H. Jasper, A. A. Ward and A. Pope, eds.). Boston: Little, Brown. pp. 13-28.

Szentágothai, J. 1970. Glomerular synapses, complex synaptic arrangements, and their operational significance. *In* The Neurosciences: Second Study Program (F. O. Schmitt, ed.-in-chief). New York: Rockefeller. pp. 427-443.

Takahashi, K. 1965. Slow and fast groups of pyramidal tract cells and their respective membrane properties. J. Neurophysiol. 28: 908-924.

Tauc, L. and H. M. Gerschenfeld. 1961. Cholinergic transmission mechanisms for both excitation and inhibition in molluscan central synapses. Nature 192: 366-367.

Thach, W. T. 1967. Somatosensory receptive fields of single units in cat cerebellar cortex. J. Neurophysiol. 30: 675-696.

Thach, W. T. 1970. Discharge of Purkinje and cerebellar nuclear neurons during rapidly alternating arm movements in the monkey. J. Neurophysiol. 31: 785-797.

Thach, W. T. 1972. Cerebellar output: properties, synthesis and uses. Brain Res. 40: 89-97.

Tömböl, T. 1967. Short neurons and their synaptic relations in the specific thalamic nuclei. Brain Res. 3: 307-326.

Tomita, T. 1965. Electrophysiological study of the mechanisms subserving color coding in the fish retina. Cold Spr. Harb. Symp. Quant. Biol. 30: 559-566.

Tomita, T. 1972. Light-induced potential and resistance changes in vertebrate photoreceptors. *In* Physiology of Photoreceptor Organs (M. G. F. Fuortes, ed.). Handbook of Sensory Physiology. vol. VII/2B. Berlin: Springer. pp. 483-511.

Towe, A. L., H. D. Patton, and T. T. Kennedy. 1964. Response properties of neurons in pericruciate cortex of cat following electrical stimulation of the appendages. Exp. Neurol. 10:325-344.

Toyoda, J., H. Nosaki and T. Tomita. 1969. Light-induced resistance changes in single photoreceptors of *Necturus* and *Gekko*. Vision Res. 9: 453-463.

Trifonov, Y. A. 1968. Study of synaptic transmission between photoreceptors and horizontal cells by means of electric stimulation of the retina. Biophysics (Moscow) 13: N 5.

Uchizono, K. 1965. Characteristics of excitatory and inhibitory synapses in the central nervous system of the cat. Nature 207: 642-643.

Uchizono, K. 1966. Excitatory and inhibitory synapses in the cat spinal cord. Jap. J. Physiol. 16: 570-575.

Uchizono, K. 1967. Synaptic organization of the Purkinje cells in the cerebellum of the cat. Exp. Brain Res. 4: 97-113.

Valdivia, O. 1971. Methods of fixation and the morphology of synaptic vesicles. J. Comp. Neur. 142: 257-273.

Valverde, F. 1965. Studies on the Pyriform Lobe. Cambridge: Harvard University Press.

Valverde, F. 1967. Apical dendritic spines of the visual cortex and light deprivation in the mouse. Exp. Brain Res. 3: 337-352.

Valverde, F. 1971. Short axon neuronal subsystems in the visual cortex of the monkey. Int. J. Neurosci. 1: 181-197.

van der Loos, H. 1960. On dendro-dendritic junctions in the cerebral cortex. *In* Structure and Function of the Cerebral Cortex D. B. Tower and J. P. Schadé, eds.). Amsterdam: Elsevier. pp. 36-42.

van der Loos, H. and E. M. Glaser. 1972. Autapses in necortex cerebri: synapses between a pyramidal cell's axon and its own dendrites. Brain Res. 48: 355-360.

van Essen, D. and J. Kelly. 1973. Correlation of cell shape and function in the visual cortex of the cat. Nature: 241: 403-405.

Van Hoesen, G. W., D. N. Pandya, and N. Butters. 1972. Cortical afferents to the entorhinal cortex of rhesus monkey. Science 175: 1471-1473.

von Euler, C. 1962. On the significance of the high zinc content in the hippocampal formation. *In* Physiologie de l'hippocampe (P. Passouant, ed.). Coll. Intern. du C.N.R.S. no. 107. Paris: Ed. du Centre Nat. Rech. Scie. pp. 135-145.

Waldeyer, W. 1891. Ueber einige neuere Forschungen im Gebiete der Anatomie des Centralnervensystems. Deutsche Med. Woch. 1352-1356.

Walshe, F. M. R. 1948. Critical Studies in Neurology. Edinburgh: Livingstone.

Watanabe, S., M. Konishi, and O. Creutzfeldt. 1966. Post-synaptic potentials in the cat's visual cortex following electrical stimulation of afferent pathways. Exp. Brain Res. 1: 272-283.

Weight, F. 1968. Cholinergic mechanisms in recurrent inhibition of motoneurons. *In* Psychopharmocology: A Review of Progress, 1957-1967. U.S. Govt. Printing Office, Washington, D.C. pp. 69-75.

Weight, F. and A. Padjen. 1973. Slow synaptic inhibition: evidence for synaptic inactivation of sodium conductance in sympathetic ganglion cells. Brain Res. 55: 219-224.

Weight, F. and J. Votova. 1970. Slow synaptic excitation in sympathetic ganglion cells: evidence for synaptic inactivation of potassium conductance. Science 170: 755-758.

Werblin, F. S. and J. E. Dowling. 1969. Organization of the retina of the mudpuppy, *Necturus maculosus*. II. Intracellular recording. J. Neurophysiol. 32: 339-355.

West, R. D. and J. E. Dowling. 1973. Synapses onto different morphological types of retinal ganglion cells. Science 178: 510-512.

Westrum, L. E. 1966. Electron microscopy of degeneration in the prepyriform cortex. J. Anat. 100: 683-685.

Westrum, L. E. 1969. Electron microscopy of degeneration in the lateral olfactory tract and plexiform layer of the prepyriform cortex of the rat. Z. Zellforsch. 98: 157-187.

Westrum, L. E. 1970. Observations on initial segments of axons in the prepyriform cortex of the rat. J. Comp. Neur. 139: 337-356.

Westrum, L. E. and T. W. Blackstad. 1962. An electron microscopic study of the stratum radiatum of the rat hippocampus (regio superior, CA 1) with particular emphasis on synaptology. J. Comp. Neur. 113: 1-42.

White, E. L. 1972. Synaptic organization in the olfactory glomerulus of the mouse. Brain Res. 37: 69-80.

White, E. L. 1973. Synaptic organization of the mammalian olfactory glomerulus: new findings including an intraspecific variation. Brain Res. 60:299-313.

Whitten, W. K. and F. H. Bronson. 1970. The role of pheromones in mammalian reproduction. *In* Communication by Chemical Signals (J. W. Johnston, Jr., D. G. Moulton, and A. Turk, eds.). Advances *in* Chemoreception, vol. 1. New York: Appleton-Century-Crofts, pp. 309-326.

Wiesendanger, M. 1973. Input from muscles and cutaneous nerves of the hand and forearm to neurones of the precentral gyrus of baboons and monkeys. J. Physiol. (London) 228: 203-220.

Willis, W. D. 1971. The case for the Renshaw cell. Brain Behav. Evol. 4: 5-52.

Wilson, V. J. and P. R. Burgess. 1962. Disinhibition in the cat spinal cord. J. Neurophysiol. 25: 392-404.

Witkovsky, P. 1971. Peripheral mechanisms of vision. Ann Rev. Physiol. 33: 257-280.

Witkovsky, P. and W. K. Stell. 1971. Gross morphology and synaptic relationships of bipolar cells in the retina of the smooth dogfish, *Mustelus canis.* Anat. Rec. 169: 456-457.

Woolsey, T. A. and H. van der Loos. 1970. The structural organization of layer IV in the somatosensory region (S 1) of mouse cerebral cortex. The description of a cortical field composed of discrete cytoarchitectonic units. Brain Res. 17: 205-242.

Yamada, E. and T. Ishikawa. 1965. The fine structure of the horizontal cells in some vertebrate retinae. Cold Spr. Harb. Symp. Quant. Biol. 30: 383-392.

Yamamoto, C. and K. Iwama. 1962. Intracellular potential recording from olfactory bulb neurones of the rabbit. Jap. J. Physiol. 38: 63-67.

Yamamoto, C., T. Yamamoto, and K. Iwama. 1963. The inhibitory system in the olfactory bulb studied by intracellular recording. J. Neurophysiol. 26:403-415.

Yokota, T., A. G. Reeves, and P. D. MacLean. 1970. Differential effects of septal and olfactory volleys on intracellular responses of hippocampal neurons in awake, sitting monkeys. J. Neurophysiol. 33: 96-107.

INDEX

Acetylcholine, 33, 101
Action potential, 37-44, 46, 49
  conduction rate, 40-43, 63
    olfactory nerve, 128
  dendritic, *see* Dendritic spike
  model, 39, 138
    mitral cell, 138
    motoneuron, 103
    olfactory pyramidal neuron, 255
  retinal axons, not in, 159, 174
  short axons, 43
  threshold, 38, 46-49, 98, 108
After potential, 38-42, 56
Alpha-gamma linkage, 94
Alpha rhythm, 233, 320
Alveus, 264, 266
Amacrine cell
  definition, 23-24
  retina, 146-176
    comparison with olfactory
      granule cell, 23-24, 117, 152
Ammon's horn, 259
Anaxonal neuron, 23-24, 117, 152,
  220, 266
Apposition, membrane, 26-28, 91, 269,
  272
Archicortex, 238, 259, 288
Autonomic nervous system, 33, 52,
  53
Avalanche conduction, 307
Axon
  branches, 21-24
  definition, 4, 7, 15, 21-26, 34, 44
    retinal neurons, 147-152, 155, 159,
      174, 176

double, cell with, 296
myelinated, 17, 20, 24, 40, 41
postsynaptic
  cerebellum, 189, 192
  neocortex, 301, 303
  olfactory cortex, 245, 246
  retina, 156, 159
  ventral horn, 90, 91
squid, 37, 39, 41
unmyelinated, 17, 20-24, 41, 42,
  114, 128
Axon collateral
  climbing fiber, 181, 196, 206
  long
    hippocampal cortex, 260
    neocortex, 293, 303, 306, 309,
      313-314
    olfactory cortex, 249-252, 257
  mossy fiber, 181, 199, 206
  primary afferent, 82
  retinal neurons, lacking in, 149,
    163
Axon hillock
  definition, 17, 20-22, 151
  impulse initiation, site of, 98, 128,
    143, 236-237
  synapses on, 189, 192, 210

Balance, 179, 182
Barrel, cortical, *see* Glomerulus,
  neocortex
Basal ganglia, 292
Basket cell
  cerebellum, 31, 181-211
    comparative aspects, 267, 270-271

Basket cell (*Cont.*)
  dentate fascia, 260, 269, 273, 283
  hippocampus, 260-281
  neocortex, 291, 295, 308, 313
  olfactory cortex, 244, 246
Betz cell, 291, 297
Bipolar cell
  neocortex, 291, 296
  retina, 146-169
    comparison with mitral cell, 135,
      160, 162
  thalamus, 220
Blood-brain barrier, 28
Boundary conditions
  branch point, 66
  closed cylinder, 62
  electrotonic spread, 62
  open cylinder, 62
Brain stem adrenergic system, 182

Cable equations, 59, 61, 69
  comparison with diffusion, 59, 327
  comparison with heat flow, 59
Capacitance, membrane, 36, 37, 47,
      60, 69
Cell body, 4, 18
Center-surround antagonism, 176
Cerebellum
  comparison with retina, 205
  comparison with thalamus, 226-228
  electric fish, 288
  simple vs. complex, 194, 196
Cerebral cortex, *see* Olfactory cortex,
      Hippocampus, Neocortex
  area of, 187
Characteristic length
  definition, 61-64
  hippocampal pyramidal neuron,
      285
  mitral primary dendrite, 136
  neocortical pyramidal neuron, 322
  olfactory glomerular tuft, 135
  retinal receptor, 171
Charging transient, 69, 71
  hippocampal pyramidal neuron,
      285
  motoneuron, 103
  neocortical pyramidal neuron, 321
Chiasm, optic, 217, 218
Chloride ion, 36, 46, 47, 49
Cilia, 25

Cingulum, 262
Cisternae, 19, 24, 25
Classical concepts, 4-9, 11, 25, 26, 32,
      33, 82, 111, 117, 127-128, 159,
      222, 224, 246
Climbing fiber, 181-212
  collaterals to deep nuclei, 199
Colliculus
  inferior, 215
  superior, 149
Colony, cortical, 317
Column, cortical, 307-309, 314-317
  comparative aspects, 127, 257, 298
Commissure, anterior, 112-114, 121,
      123, 142
Compartmental model
  equivalent cylinder, 70-74
  metabolic studies, 71
Complex
  amygdaloid, 239
  inferior olivary, 182, 202-203, 279
  synaptic, *see* Synaptic complex
  ventrobasal, 214-225, 314
Complex cell, 315, 316
Computer, analogy with, 169, 205,
      210
Conditional response, 319
Conditional stimuli, 212-213, 278
Conductance, 36
  dendritic input, 65, 67
  somatic input, 65
Consciousness, 319
Contrast enhancement, 231
Convergence
  cerebellum, 187-188, 197-198
  hippocampus, 266
  neocortex, 300, 312
  olfactory bulb, 118
  olfactory cortex, 244
  retina, 153-154
  thalamus, 221-222
  ventral horn, 82, 88
Convolutions
  cerebellum, 180
  neocortex, 288
Core conductor, 58, 76
Corpus callosum, 292
Cortex
  agranular, 295
  definition, 127, 163-164, 195, 257-
      258, 309, 328
  granular, 295

Cyclic AMP, 52, 202
Cytoarchitectonics, 297-299

Dale's Law (Principle), 29-30, 53-54,
    101, 201, 245
Dark current, in retinal receptors,
    166, 170-172
De-afferentation, 57, 208, 278
Dendrite
    active vs. passive, 44, *see also*
        Dendritic spike
    apical
        hippocampus, 263
        neocortex, 293
        olfactory cortex, 241
    basal
        hippocampus, 264
        neocortex, 293
        olfactory cortex, 242
    branching, 66-69, 71
    definition, 4, 7, 15, 22-26, 34, 118,
        152, 174
    diameters, 63-66, 71
    functional properties, 43
    lengths, 62-63, 71
    presynaptic
        neocortex, 301, 321, 325
        olfactory bulb, 117
        retina, 156
        thalamus, 223, 236-237
    primary, 115
    secondary, 115
    topography, 177, 235
Dendritic dominance, 65, 321
Dendritic spike, 44
    amacrine cell, 176
    hippocampal pyramidal neuron,
        287
    mitral cell, 137-139
    motoneuron, 106
    muscle spindle, 209
    neocortical neuron, 325
    olfactory granule cell, 141, 142
    Purkinje cell, 207-210
    significance, 209-210
    spine, 141
Dendritic spine, 16, 25
    dentate granule neuron, 267, 271
    hippocampal intrinsic neuron, 265
    hippocampal pyramidal neuron,
        260, 264, 269, 271
    modifiability, 212, 326-328

neocortical intrinsic neuron, 295-
    296
neocortical pyramidal neuron, 293,
    300, 301, 325-328
olfactory granule cell, 117-123, 131,
    132, 140-142, 183
olfactory pyramidal neuron, 242,
    245, 246
Purkinje cell, 183, 189, 191, 197,
    200, 211-213
thalamic intrinsic neuron, 237
thalamic relay neuron, 219
Density, neuronal
    cerebellum, 300
    neocortex, 300
Dentate fascia, 259-284
    comparative aspects, 277
Depression, synaptic, 56, 282, 284
Desmosome, 27, 189, 190, 227
Diffusion, in spine, 327-328
Digestive tract, uptake, 56
Dipole, potential, 253
Divergence, 82, 85
    cerebellum, 187-188, 197-198, 227
    hippocampus, 267, 281
    thalamus, 222, 227
Dolphin, lacks olfactory bulbs, 259
Dorsal root ganglion cell, 4, 5, 22, 23,
    79-94, 215
Dynamic differentiation, of cerebel-
    lar nuclear cell output, 207
Dynamic polarization, 6

Electrical stimulation, of cortex, 317
Electroencephalographic (EEG)
    waves,
    hippocampus, 282
    neocortex, 320
    olfactory bulb, 134
    olfactory cortex, 253
    thalamus, 233
Electrotonic coupling, 157, 175
Electrotonic currents, 50, 60-67
Electrotonic length, 60-68, 70, 71,
    see also Electrotonic model
    periglomerular cell, 143-144
    thalamic neuron, 234
Electrotonic model, 69
    mitral primary dendrites, 136-137
    mitral secondary dendrites, 138
    motoneuron, 102-106
    neocortical pyramidal neuron, 323

Electrotonic model (*Cont.*)
  olfactory glomerular tuft, 135-136
  olfactory granule cell, 140, 141
  olfactory pyramidal neuron, 254,
    255
  retinal bipolar cell, 173
  retinal receptor, 170-172
Electrotonic potential, 48, 58-69
  dependence on dendritic branch-
    ing, 66-70
  dependence on dendritic diameter,
    63-66, 68, 70
  dependence on dendritic length,
    62-63, 70
  hippocampal pyramidal neuron,
    285
  horizontal cell, retina, 175
  mitral primary dendrite, 136-137
  mitral secondary dendrites, 138,
    139
  motoneuron, 103-108
  neocortical pyramidal neuron, 324
  olfactory glomerular tuft, 136
  olfactory granule cell, 140
  olfactory pyramidal cell, 255
  as passive potential, 61
  retinal receptor, 171, 172
Electrotonus
  definition, 58
  dendritic, 44, 58-76
  steady-state, 59-69
  transient, 69-76
Encephalization of nervous control,
    176
Endoplasmic reticulum, 17, 25
  rough, 17-19
  smooth, 17-21
End-plate potential, 51, 59
  miniature, 50
Engineering, systems, 253
Entorhinal cortex, 262, 267, 277, 278
  transitional, 239
Ephapse, 26, 272
Epilepsy, *see* Seizure
Epileptic aggregate, 283
Epileptic neuron, 283
Equivalent circuit, for neuronal
    membrane, 36
Equivalent cylinder, 66, 67
  compartmental model, 70-74
  mitral secondary dendrites, 137,
    138

motoneuron, 102-106
olfactory dendritic tuft, 135-136
olfactory granule cell, 139, 140
olfactory pyramidal neuron, 254,
    255
periglomerular cell, 143
spread of potential in, 68
thalamic neuron, 234
Evoked potential, *see* Extracellular
    potentials, summed
Evolution
  of central nervous system, 288
  of cortex, 238
  of hippocampus, 259
  of intelligence, 306
  of neocortex, 288-289
  of retina, 145
Excitability cycle, of membrane, 40
Excitation, synaptic, definition, 48
Excitatory-inhibitory sequence
  cerebellum, 200, 203
  dentate fascia, 283
  hippocampus, 280, 282
  neocortex, 312
  olfactory bulb, 129
  olfactory cortex, 250
  thalamus, 229-230, 235
Extracellular potentials, summed
  hippocampus, 282
  neocortex, 320
  olfactory bulb, 129-130, 134, 140,
    142
  olfactory cortex, 251, 253
  potential divider effect, 130

Facilitation, synaptic, 56, 99, 282,
    284, 319, 325
Feedback inhibition, *see* Recurrent
    inhibition
Fiber
  association
    dentate fascia, 271
    hippocampus, 264
    neocortex, 292, 294
    olfactory cortex, 241, 243
  centrifugal, olfactory bulb, 113-
    114, 119-121, 126, 136
    compared to ventral horn, 114
  centrifugal, retina, 148
  climbing, *see* Climbing fiber
  commissural, 263, 271, 273, 292, 294
  corticospinal, 97, 107

corticothalamic, 217-229
flexor reflex afferent, 93-99
group Ia afferent, excitatory, 93-
      108, 157, 318
group Ia afferent, inhibitory,
      93-107
group Ib afferent, 95
group II afferent, 95
mossy, *see* Mossy fiber
nonspecific sensory afferent, 290-
      292, 305, 306, 311, 319, 323, 324
parallel, 16, 22, 26, 181-213, 248, 272
peripheral afferent, 80-82, 92, 93
   compared to olfactory nerve, 114
projection, neocortical, 294
specific sensory afferent, 290-292,
      304-306
Field potential, *see* Extracellular
   potentials, summed
Fimbria, 264
Final common path, 6, 79
Firing frequency
   cerebellar neurons, 204-205
   deep cerebellar neurons, 206
   ganglion cells, retina, 168
   hippocampal neurons, 280
   motoneurons, 105
   neocortical neurons, 318
   olfactory bulb neurons, 144
   Renshaw units, 100, 101
   thalamic neurons, 230
Fornix, 259-282
Fractionation of dendritic field, 107
Freeze-fracture, 16
Functional complexity
   dependent on synaptic connec-
      tions, 154, 306
   large neurons, 297
   small neurons, 306
Functional unit
   amacrine cell, 175
   bipolar cell, 162
   bipolar-ganglion cell, 164
   classical concept, 8
   dendritic, 8
   ganglion cell, 163
   Golgi cell, 202
   horizontal cell, 174
   mitral dendrites, 134, 162, 163
   multineuronal, 8, 163, 164, 211
   neocortex, 307, 315, 328
   olfactory glomerulus, 127

synaptic, 8
thalamic neurons, 236-237
Functions of neurons, 35
Fusiform cell, neocortex, 296

Ganglion
   spinal, 215
   sympathetic, 55
Ganglion cell, retina, 146-177, 215,
      317
   axon collaterals, lacking in, 149
   axon to optic nerve, 215
Gemmule, 25, 117, 120
Genetic differences, in synaptic connectivity, 119-120
Geometry
   cerebellum, 180, 183, 198
   hippocampus, 274
   olfactory cortex, 248
Glia (neuroglia), 42, 122, 150, 189,
      191, 222, 223, 225, 320
   Bergman, 192
Glomerulus
   definition, 191, 299
   macroglomerulus, 191, 202, 223,
      228, 299, 308
   microglomerulus, 191, 223
   neocortex (barrel), 202, 299, 308-
      309
   olfactory, 113-144, 191, 202, 228,
      234, 245, 299
   synaptic, 191
      cerebellum, 190-191, 222, 225-228
      retina, 158
      thalamus, 191, 222-231
      ventral horn, 91
Glycogen granules, 150
Golgi body, 17, 19, 150
Golgi cell, cerebellum, 181-204
   axonal arborization, 186, 281
Golgi stain, 6, 11, 17, 34, 150, 152, 155,
      177, 221, 246, 303, 323
Golgi type II neuron, 220, *see also*
   Short-axon cell
Granule cell
   cerebellum, 181-204
      comparative aspects, 22, 96, 226-
         228, 248
      numbers, 188, 299
      packing density, 300
      parallel fibers, gives rise to, 185
   definition, 117, 185, 241, 267, 295

Granule cell (*Cont.*)
  dentate fascia, 96, 241, 260-284
  neocortex, 291, 295
    comparative aspects, 96, 306
    packing density, 300
  olfactory bulb, 113-142, 239
    comparative aspects, 23, 32, 152,
      160, 161, 175, 233, 237, 245

Hand, control of, 84, 85, 94, 99, 318
Heart, inhibition, by vagus, 53, 54
Heat production, of unmyelinated
  axons, 42
Horizontal cell, retina, 146-175
Hypercomplex cell, 315-317
Hypothalamus, 33, 126, 216, 239,
  243, 259, 264-265, 279

Impulse, *see* Action potential
Inactivation response (burst), 202,
  279
Inferior olive, *see* Complex, inferior
  olivary
Inhibition
  afferent, 231
  axon hillock, 109
  criteria for, 200
  definition, 48
  dendritic, 107-109, 138, 210-211
  disinhibition
    cerebellum, 201, 204
    olfactory bulb, 144
    ventral horn, 96, 102, 109, 110
  distal, 109, 110
  feedback, *see* Recurrent inhibition
  feedforward
    cerebellum, 196, 201, 203
    neocortex, 308
    olfactory bulb, 131-134
    retina, 169
    thalamus, 230, 232
    ventral horn, 96, 109, 110
  lateral, 132, 141, 231
  presynaptic, 91, 109, 110, 167, 231,
    252
  proximal, 109, 110
  reciprocal, 94
  recurrent, *see* Recurrent inhibition
  remote, 109, 110
  self-, 132, 141
  soma, 109, 302
  surround, 102

Inhibitory phasing of neuronal discharge, 233, 252-253, 320
Initial segment, 17, 20-22
  site of impulse initiation, 98, 104-
    107
  synapses onto, 75, 189, 245, 246
Input fiber
  comparative aspects, 114, 147, 182-
    183, 241, 297
  definition, 4-5
Input-output relations
  cerebellum, 198, 205-206, 211-212
  electrotonic factors in, 73, 75-76
  hippocampus, 266-267
  neocortical spines, 328
  olfactory bulb, 123, 124-125, 133,
    142
  olfactory cortex, 244
  retina, 153, 172, 174
  thalamus, 221, 236-237
  ventral horn, 88, 102
Interneuron, 4, 85, *see also* Intrinsic
  neuron
Intrinsic neuron
  definition, 4-5, 23, 86-87, 243-244,
    294
  impulse in, 43
Invertebrate neurons, 23, 45, 54, 56,
  57, 297
Ionic current flow
  during action potential, 38
  during synaptic potential, 46, 49
Ionic pump, 37, 42
  electrogenic, 42, 56
Island, synaptic, *see* Glomerulus,
  synaptic

Junction
  gap, 27, 28, 45, 91, 157, 203
  tight, 27, 28, 91
Juxtaposition, membrane, 26, 27, 90

Kirkhoff's laws, 61

Lamination
  cerebellum, 181
  hippocampus, 260, 262, 276
  lateral geniculate nucleus, 218
  neocortex, 291, 298, 309
  olfactory bulb, 113, 127
  olfactory cortex, 240
  retina, 146, 163

ventral horn, 80
Learning, synaptic basis for, 57, 212, 261, 278, 282, 284, 323, 327
Leech, 56
Light, threshold for perception of, 170
Local neuron, *see* Intrinsic neuron
Local region
  cerebellum, 195
  definition, 4
  neocortex, 294
  olfactory cortex, 243-244
  ventral horn, 84, 88
Locus coeruleus, 182
Lysozyme, 19

Mammillary body, 264
Mauthner cell, 192
Membrane patch model, 60, 72
Membrane potential, 35-37, 42, 49
  gradient in retinal receptor, 172
  low values in retinal neurons, 164-168
  resting value, 61
Membrane turnover, Golgi body, 19
Memory, synaptic basis for, 57, 212, 261, 278, 282, 284, 323
Metabolic energy, mitochondria supply, 19
Metabolic unity of neuron (Dale's Law), 53-54
Metabolism, in unmyelinated axons, 42
Micron, 14
Microtubule, 17-25
Mitochondria, 17-25, 172
Mitral cell, 32, 113-144, 152, 160, 161, 175, 236, 239
  axons of (lateral olfactory tract), 239, 240
  compared with motoneuron, 115
  compared with retinal ganglion cell, 162
  compared with thalamic neuron, 233-235
Mossy fiber
  cerebellum, 31, 181-206
    collaterals to deep nuclei, 199
  hippocampus, 260-281
Mossy terminal
  cerebellum, 182, 189, 190
    rosette, 190, 196, 202

hippocampus, 269, 270
Motoneuron, 4, 5, 43, 65, 66, 79-110, 116, 152, 157, 204-205, 234, 318
  alpha, 80-105, 152
  axons of (ventral root), 87
  chromatolytic, 106, 287
  flexor and extensor, 93, 94, 317
  gamma, 83-105, 115, 152
  somatotopical organization, 84
  surface area, 92
Motor unit, 85
Movement
  control, 289, 297, 318
  coordination, 179, 212
  initiation, 212, 318
Multivesicular body, 19
Muscle spindle, 83, 91-98, 105, 179, 209, 311, 318-319
Muscle tone, 105, 179
Myelin
  around axons, 21, 24
  around cell bodies, 122
  around dendrites, 122, 137

Necturus (mudpuppy), 154, 165
Neocortex
  association area, 289
  layers of, 291
  motor area, 289, 295-297, 300, 304-305, 308-314, 317-319
  organizational principles, 257-258, 275, 297, 307, 309
  sensory areas, 289, 304, 305
  somatosensory area, 299, 314
  thalamic relations, 214-219
  visual area, 232, 290, 295, 296, 300, 315
Nerve
  auditory, 154
  deep radial, 311
  olfactory, 111-143, 152, 185, 191, 272
  optic, 148-154, 217-222
  superficial radial, 311
  vagus, 53, 54
  vestibular, 182
Nest, synaptic, *see* Glomerulus, synaptic
Neurite, 23
Neurofilament, 17, 19-25
Neuromuscular junction, 28, 31, 50-56, 59, 101, 130

Neuron, definition of, 150
Neuron doctrine, functional concepts, 6, 25, 32, 111, 159, 224, 246
Neurosecretory cell, 33, 34
Nissl substance, 17, 19, 20, 150, 272
Node of Ranvier, 21, 40, 41
Nucleus
   anterior olfactory, 112, 113, 121, 142, 239
   brain stem, 292
   cochlear, 30, 215
   deep cerebellar, 184-207
   definition, 195, 258
   dentate, 184
   diagonal band, 113, 114, 121, 123, 239
   dorsal column, 4, 5, 215, 314
   intermediate (Cajal), 82, 94, 110
   lateral geniculate, 149, 214-237, 256, 292, 315, 316, 326
   lateral lemniscus, 215
   medial geniculate, 214-224
   motoneurons arranged in, 81
   nonspecific thalamic, 216, 292, 294, 311, 319, 323
   olivary, inferior, *see* Complex, inferior olivary
   olivary, superior, 215
   pontine, 182
   red, 184
   reticular, 82, 182, 184
   specific thalamic, 215, 319, 322
   ventrolateral, thalamus, 184
   vestibular, 83, 182
Nucleus, of cell, 17, 18

Occluding junction, 27
Ohm's Law, 67
Olfactory axon, 26, 41, *see also* Olfactory nerve
Olfactory bulb, comparison with retina, 160
Olfactory stimulation, 127, 134, 136, 143, 256
Olfactory tubercle, 239
Oxygen consumption, in unmyelinated axon, 42

Pacemaker potential, *see* Prepotential
Palaeocortex, 238, 288
Pallium, 288

Parcellation
   hippocampus, 275, 281
   neocortex, 309
Paroxysmal depolarizing shift, 283, 320-321
Particles
   membrane, 16
   ribonucleic, 17-25
Parts of the neuron, 17-26, 34, 35, 44
Passive potential, 61, *see also* Electrotonic potential
Pathway, *see* System
   alvear, 260-272
   perforant, 260-286
Periglomerular cell, 113-144, 160, 161, 237
Perikaryon, 18, 20
Pheromone, 239
Photon, 170-172
Pituitary, 33
Plasma membrane, 15
Polymorphic cell, *see* Stellate cell
Post-inhibitory exaltation, 234
Posture, 179
Potassium ion, 36-49, 55
   depolarization, long-lasting, 320
Potential divider, in extracellular recording, 130
Potentiation, synaptic, 325
Prepotential, 38, 40
   fast, 285-287, 325
Prepyriform cortex, *see* Olfactory cortex
Presynaptic control, 163, *see also* Inhibition, presynaptic
Presynaptic dendrite, 25, *see also* Dendrite, presynaptic
Principal neuron, definition, 4, 5, 22, 23, 87, 195, 243, 276, 277, 294
Projection neuron, *see* Principal neuron
Propriospinal neuron, 85-87, 102
Protein synthesis
   Nissl bodies sites of, 19
   ribosomes sites of, 18
Protoplasmic prolongations (dendrite), 23
Purkinje cell, 16, 20, 181-213, 248, 267, 271, 279, 287
   inhibitory output, 206
Pyramidal neuron
   definition, 241-242, 268, 310

dentate fascia, 268
hippocampus, 260-287
neocortex, 291-316
olfactory cortex, 241-258

Quantum
of light, 170
of transmitter substance, 50, 51, 108

Rebound excitation, 280
Receptor
muscle stretch, 82, 95
olfactory, 111-113, 118
pain, 82
pressure, 82
retinal, 31, 69, 166
temperature, 82
touch, 82
Receptor potential, retinal receptors, 166
Rectification, of electrical synapse, 45
Recurrent collateral, 22
dentate granule cell, 267, 283
hippocampal pyramidal neuron, 260, 264, 273, 280
Schaffer collaterals, 260, 264, 271-275
inferior olive neuron, 203
mitral cell, 113, 123, 246
motoneuron, 80, 93, 100
neocortical pyramidal neuron, 291-294, 303, 308, 313, 323
olfactory pyramidal neuron, 240, 247, 249, 252, 257
Purkinje cell, 181, 192-195, 201, 203
retinal neurons, lacking in, 149, 163
thalamic relay neuron, 218, 228, 230
primate, lacking in, 219, 230
Recurrent excitation
cerebellum, not present in, 204
hippocampus, 273, 280
neocortex, 305-308, 313-314, 317, 320, 323
olfactory cortex, 250-253
spinal cord, 99
Recurrent inhibition
axon collateral (direct), 194
axon collateral (Renshaw)
cerebellum, 196

deep cerebellar nuclei, 194
denate fascia, 283
hippocampus, 280-283
neocortex, 305, 308, 313, 323
olfactory bulb, 132
olfactory cortex, 249, 252
thalamus, 230-233
ventral horn, 100-102, 107-110
dendrodendritic
neocortex, 313
olfactory bulb, 131-133
retina, 169, 175-176
thalamus, 232-233
Re-excitation, *see* Recurrent excitation
Reflex, muscle, 40
Reflex, stretch, 95
Reflex arc, 4, 5, 79, 318
Refractory period, 40
Relay neuron, *see* Principal neuron
Renshaw cell, 93-110, 132
problem of identification, 101-102
Repetitive stimulation, after-effects of
dentate fascia, 284
leech neurons, 56
neuromuscular junction, 56
unmyelinated axons, 42
Resistance
internal, 36, 58, 60, 63
membrane, 36, 58, 60, 63
specific internal, 63-67
specific membrane, 63-68
hippocampal pyramidal neuron, 284
motoneuron, 103-104
neocortical pyramidal neuron, 321, 325
retinal receptor, 171
spine stem, 211, 327
whole neuron (input), 65, 69
hippocampal pyramidal neuron, 284
horizontal cell, retina, 175
motoneuron, 103, 105, 108
neocortical pyramidal neuron, 321, 323
Reticular formation, 216, 218
Retina
comparison with olfactory bulb, 160
simple vs. complex, 145-177, 315

Retinal receptor, 31, 69, 146-172
  cone, 146
  cone pedicle, 156
  rod, 146
  rod spherule, 156
Reverberating circuits, 307
Rhinencephalon, 259-260
Rhythmic activity
  hippocampus, 261, 279, 281-283
  neocortex, 319-321
  olfactory bulb, 133-134
  olfactory cortex, 252-253
  thalamus, 230, 232-233
Ribonucleic particles, 17, *see also*
     Particles, ribonucleic
Ribosomes, 17, *see* Particles, ribo-
     nucleic
RNA synthesis, 56
Rosette, *see* Mossy terminal, cere-
     bellum

Safety factor, 40
Saltatory conduction, 40, 287
  microsaltatory, 41
  pseudosaltatory, 209
Scale
  distance, 12
  time, 12
Scaling principle
  motoneuron, 105
  neocortical pyramidal neuron, 322
  olfactory glomerular tuft, 136
  periglomerular cell, 143-144
  thalamic neuron, 234, 235
Schaffer collateral, *see* Recurrent
     collateral, hippocampal pyra-
     midal neuron
Schwann cell, 21, 122
Seizure, 261, 282-283, 320-321
Sense of effort, 319
Sensory deprivation, 326
Sensory processing
  neocortex, 314-317
  olfactory bulb, 126, 145
  olfactory cortex, 256-257
  retina, 177
Septate junction, 27
Septum (septal region), 239, 259-279
Serial reconstruction, 34, 120, 155,
     158
Servo loop, cortical, 318
Sheet, dendritic, 25
Short-axon cell

cerebellum, 181, 186
definition, 86-87, 116-117, 186-187,
     220, 244
dentate fascia, 268
hippocampus, 265
impulses in, 43
olfactory bulb, 113, 116-118
olfactory cortex, 244
thalamus, 144, 220-221, 230, 236-
     237
ventral horn, 86-87, 101
Simple cell, 315, 316
Size principle, 106, 144, 204
Snail (*Aplysia*), 52, 56
Sodium ion, 36-47, 55, 166
  density of channels, 41
  flow in retinal rods, 166, 167
  movement without conductance
     increase, 55
Soma, 4, 18
Spatio-temporal factors, 198, 211
Spike, in Purkinje cell
  complex, 200, 202-204, 279
  simple, 200
Spinal cord, dorsal horn, 91, 94, 110
  dendrodendritic synapses in, 91
Spine apparatus, 17, 25, 269, 301
  axon, 246
  axonal initial segment, 245, 246
Spinule, *see* Synapses, types of
S potential, in horizontal cell, 166,
     174
Stain, intracellular, 17, 34, 44, 85,
     102, 104, 150, 164, 209, 316
Stellate cell
  cerebellum 181-211
  neocortex, 295, 316
  olfactory cortex, 240-252
Stretch receptor, crayfish, 75, 105,
     106-107
Structure-function relations, 44, 53,
     142, 147, 169, 177-178, 235, 241
Subiculum, 262, 265
Summation, synaptic, 197
  spatial, 76
  temporal, 99, 285
Synapses, arrangements of
  axoaxonic, 32, 90, 189, 193, 245,
     303
  axodendritic, 27, 32, 90, 117, 189,
     223, 245, 269, 301
  axosomatic, 31, 90, 193, 245, 269,
     301

dendroaxonic, 159
dendrodendritic, 27, 32, 91, 101,
    119, 156, 223, 230, 237, 301, 326
  mechanism, 50, 73, 75, 132, 175,
      324
dendrosomatic, 121
dyad, 158
  reciprocal, 27, 32, 117, 156
    arrangement, 32
    pair, 32, 117, 156
  serial, 27, 32, 90, 119
  somatodendritic, 121
  triad, *see* Triad, synaptic
  triad, retina, 157
Synapses, modifiability of, 57, 212,
    326-328
Synapses, types of
  adrenergic, 33, 52-54, 182
  autapse, 303, 308
  chemical, 28-34, 46-57
    type I, 27, 29
    type II, 27, 29
  cholinergic, 33, 52-54
  conventional, 32, 157, 158
  crossing over, 191
  electrical, 28, 45-46, 157, 175
  en marron, 190
  en passage (passant), 22, 191, 192,
      270
  multifunctional (multiplex), 126,
      163, 173
  nonusual, 32
  ribbon, 157, 158, 160-162, 167
  simple, 30-34, 91, 159
  specialized, 27, 31-34, 50, 157, 159
  spinule, 31, 269, 271
  superficial, 157
  unconventional, 32, 33
Synaptic cleft, 16, 29, 50, 51
Synaptic complex, *see also* Glo-
    merulus, synaptic
  cerebellum, 191
  definition, 191
  olfactory bulb, 120, 122
  retina, 156
  ventral horn, 91
Synaptic delay, 51, 97-98, 101
Synaptic glomerulus, *see* Glomerulus,
    synaptic
Synaptic integration, 48-50, 74, 97, 98
  electrotonic properties, 73-76
  motoneuron, 97-99, 106-108
  neocortical pyramidal neuron, 322-

325
  nonlinear properties, 48-49
  olfactory pyramidal neuron, 255-
      256
Synaptic lamella (ribbon), 157
Synaptic potency, 99, 198, 203
  changes in, 57, 282, 284, 320, 328
Synaptic potential
  electrotonic properties, 71-74
  excitatory postsynaptic (EPSP),
      46-48, 55, 56
    cerebellum, 200
    dentate fascia, 283-284
    hippocampus, 278
    neocortex, 311, 323
    olfactory bulb, 129, 133, 140
    olfactory cortex, 251, 255
    thalamus, 230
    ventral horn, 97, 100, 103, 107
  graded, 47
  inhibitory postsynaptic (IPSP),
      46-49, 55, 56
    cerebellum, 200
    deep cerebellar nuclei, 206
    dentate fascia, 283-284
    hippocampus, 278
    neocortex, 311, 323
    olfactory bulb, 129, 133, 138
    olfactory cortex, 251, 255
    thalamus, 230
    ventral horn, 97, 100, 103, 107
  interactions, 73-75, 97-99, *see also*
      Synaptic integration
  leech neurons, 56
  linear, 47
  local vs. distant, 73-76
  long-lasting, 130-131, 201, 213, 281,
      282, 319
  miniature, 108
  retina, 165
  shape index, 72, 104, 322
  sympathetic ganglion, 55
System
  adrenergic, 182
  auditory, 215, 261
  corticothalamic, 320
  limbic, 126, 239, 261-265
  motor, 4, 5, 93-95, 317-319
  olfactory, 239, 248, 261, 278
  reticular activating, 292
  somatosensory, 215, 261
  thalamocortical, 214
  visual, 215, 261

Tectum, 149
Telodendria (axon terminals), 25
Temporal lobe, 262
Tendon organ, 95
Terminal
    axonal, 17, 22-23, 25, 50
    definition, 25-26
    dendritic, 17, 23, 25
    form determined by postsynaptic
        site, 183
    giant (neuromuscular), 50
    as input-output unit, 236-237
    large size, 155, 158, 190
    mossy, *see* Mossy terminal
    postsynaptic, 46, 51, 52
    presynaptic, 46, 50-52
    simple, 31-34, 91, 159
    specialized, 27, 31-34, 159, 189
Thalamic relay neuron, 135, 136, 217-
        236
    axon collaterals lacking in, 220-221
Thalamus, *see also* Nucleus
    anterior, 264-265
    comparison with cerebellar granule
        cells, 196
    comparison with olfactory glo-
        merulus, 136
    sensory nuclei, fraction of, 216
Theta rhythm, 282
Time constant, 69
    charging, 69
        hippocampal pyramidal neuron,
            285
        motoneuron, 103
        neocortical pyramidal neuron,
            321
    equalizing, 72
    membrane, 71
        hippocampal pyramidal neuron,
            285
        motoneuron, 103
    whole neuron, 69
Tract
    corticospinal, 5, 82, 97, 99, 310
    lateral olfactory, 112, 113, 239, 240
    pyramidal, 5, 310
    spino-cerebellar, 182
Transmitter substances, 50-55
    acetylcholine, 33, 51-54
    amino acids, 33, 54, 55
    aspartic acid, 54, 167

cyclic AMP, 52, 202
diversity of, 53
gamma-aminobutyric acid
    (GABA), 55
glutamic acid, 54, 55
glycine, 55
norepinephrine, 33, 52
polypeptide hormone, 34
release by membrane depolariza-
    tion, 166, 236-237
Transport, intracellular
    retinal receptors, 172
    role of microtubules, 19
Triad
    retinal, 155, 157
    synaptic, 5
        cerebellum, 195-197, 199
        hippocampus, 274
        neocortex, 305
        olfactory bulb, 124-126
        olfactory cortex, 248
        retina, 162
        thalamus, 225-228
        ventral horn, 96
Trigger zone, 286
Tufted cell, 115-127, 152

Unconditional response, 319
Unconditional stimuli, 212, 213, 278
Undercoating
    initial segment, 20
    node of Ranvier, 21
Unit membrane, 18

Vesicle, synaptic, 16-25, 28, 31-34, 52,
        192
    dense core, 33-34, 269
    flattened, 29
    large, 33-34
    medium-size, 33
    quantum, relation to, 50, 108
    small, 33
    spherical, 29
Vestibular canals, 179, 182
Vision
    acuity, 153, 177, 217
    dominant role of, 154

Zinc, in hippocampus, 270
Zonulae adherens, 27